INTR DUCTION TO

General
Relativity

SOLUTIONS TO PROBLEMS

INTRODUCTION TO

General Relativity

SOLUTIONS TO PROBLEMS

JOHN DIRK WALECKA

College of William and Mary, USA

World Scientific

NEW JERSEY · LONDON · SINGAPORE · BEIJING · SHANGHAI · HONG KONG · TAIPEI · CHENNAI · TOKYO

Published by

World Scientific Publishing Co. Pte. Ltd.

5 Toh Tuck Link, Singapore 596224

USA office: 27 Warren Street, Suite 401-402, Hackensack, NJ 07601

UK office: 57 Shelton Street, Covent Garden, London WC2H 9HE

Library of Congress Cataloging-in-Publication Data
Names: Walecka, John Dirk, 1932–   author.
Title: Introduction to general relativity : solutions to problems / John Dirk Walecka,
    College of William and Mary, USA.
Description: Singapore ; Hackensack, NJ : World Scientific, [2017] |
    Includes bibliographical references and index.
Identifiers: LCCN 2017023318| ISBN 9789813227699 (pbk. ; alk. paper) |
    ISBN 9813227699 (pbk. ; alk. paper)
Subjects: LCSH: General relativity (Physics)--Mathematics--Problems, exercises, etc.
Classification: LCC QC173.6 .W355 2017 | DDC 530.11--dc23
LC record available at https://lccn.loc.gov/2017023318

British Library Cataloguing-in-Publication Data
A catalogue record for this book is available from the British Library.

Printed in Singapore

Preface

Eleven years after Einstein's theory of special relativity completely changed our understanding of the relationship between space and time [Einstein (1905)], his theory of general relativity revolutionized our understanding of how mass and energy change the underlying space-time structure of the physical universe [Einstein (1916)].

Some time ago I had to learn general relativity in connection with my research. I had great difficulty listening to the experts and understanding what they were doing. I decided I had to teach the subject to myself, and I proceeded to do so. Much later, I converted my notes to a semester course at William and Mary, which was offered three times. It was aimed at physics graduate students, but many undergraduates participated, and excelled. The students seemed to learn from it and enjoy it, and the outcome was very satisfying. I decided that I would convert these lectures into a book entitled *Introduction to General Relativity*, which was subsequently published by World Scientific Publishing Company [Walecka (2007)].

General relativity is a difficult subject for two reasons. The first is that the math is unfamiliar to most physics students, and the second is that since the four-dimensional coordinate system has no intrinsic meaning, it is difficult to get at the physical interpretation and physical consequences of any result. The goals of the text *Introduction to General Relativity* are as follows:

- The book is aimed at physics graduate students and advanced undergraduates. Only a working knowledge of classical lagrangian mechanics is assumed,[1] although an acquaintance with special relativity will make the book more meaningful. Within this framework,

---

[1] As presented, for example, in [Fetter and Walecka (2003)].

the material is self-contained;

- The necessary mathematics is developed within the context of a concrete physical problem — that of a point mass constrained to move without friction on an arbitrarily shaped two-dimensional surface;

- A strong emphasis is placed on the physical interpretation and physical consequences in all applications;

- The book is not meant to be a weighty tome for experts (indeed, I could not write one), but, in my opinion, most of the interesting applications of GR are covered;

- The final "Special Topics" chapter takes the reader up to a few areas of active research.

In order to enhance and extend the coverage, some 108 problems have been included in that book. For the most part, the problems are not difficult, and the steps are clearly laid out. Although nothing replaces a direct confrontation of the problems, the present book provides the solutions to those problems. I believe there is interesting, useful material in this *solutions manual*, and it should be of benefit to both students and teachers.

At the end of the course at William and Mary, students selected a topic of current interest, wrote a term paper, and then gave a talk to the class in the format of a contributed session of an American Physical Society meeting. I was amazed and pleased with the level at which the students were able to perform. It is my sincere hope that this text and solution manual will provide useful background for other young people and aid them in their exploration of the many fascinating modern developments based on Einstein's theory of general relativity.

I would like to thank Dr. K. K. Phua, Executive Chairman of World Scientific Publishing Company, and my editor Ms. Lakshmi Narayanan, for their help and support on this project.

Williamsburg, Virginia   *John Dirk Walecka*
April 6, 2017   *Governor's Distinguished CEBAF*
         Professor of Physics, Emeritus
         College of William and Mary

Contents

Chapter 1

Introduction

There are no problems associated with the Introduction.

Chapter 2

Particle on a Two-Dimensional Surface

Problem 2.1 Consider two parallel flat surfaces interconnected through a smooth circular cylindrical tube to make a single unified surface. Give a qualitative discussion of the particle orbits and geodesics on the unified surface.

Solution to Problem 2.1

The easiest way to track the geodesics is to imagine a string connecting two points on the surface (which we assume is smooth), and then just pull the string tight. We sketch three geodesics in Fig. 2.1: one is unperturbed on the lower surface, one starts on the upper surface, dips into the cylinder, and then returns on the upper surface, and one starts on the upper surface, dives down the cylinder, and returns on the lower surface. There are also, of course, orbits with different winding numbers around the cylinder.[1]

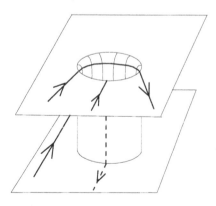

Fig. 2.1   Sketch of three geodesics on the given two-sided surface (see text).

---

[1] Readers are urged to envision a few of these.

Chapter 3

Curvilinear Coordinate Systems

Problem 3.1 Given the definition of $\delta^i{}_j$ in Eq. (2.13), the fact that the metric raises and lowers indices, and Eq. (2.20), show $\delta_i{}^j = \delta^i{}_j$.

Solution to Problem 3.1

The basis vectors in the reciprocal basis satisfy [see Eq. (2.13)]

$$\mathbf{e}^i \cdot \mathbf{e}_j = \delta^i{}_j$$
$$= 1 \qquad ; \text{ if } i = j$$
$$= 0 \qquad \text{ if } i \neq j$$

The metric in the reciprocal basis is defined in Eqs. (2.17)

$$\mathbf{e}^i \cdot \mathbf{e}^j \equiv g^{ij} = g^{ji}$$

It is also symmetric, and satisfies Eq. (2.20)

$$\sum_{l=1}^{2} g^{il} g_{lj} = \delta^i{}_j$$

Since the metric raises and lowers indices, one has

$$\delta_i{}^j = \sum_{i'} \sum_{j'} g_{ii'}\, \delta^{i'}{}_{j'}\, g^{j'j} = \sum_{i'} g^{ji'} g_{i'i}$$
$$= \delta^j{}_i = \delta^i{}_j$$

where the last equality follows from the definition of $\delta^i{}_j$.

Problem 3.2 Suppose a contraction $v_i w^i$ is invariant under coordinate transformations $v_i w^i = \bar{v}_i \bar{w}^i$ for an *arbitrary* vector v_i, which transforms as $\bar{v}_i = \bar{a}_i{}^j v_j$. Show that w^i must then also be a vector transforming as $\bar{w}^i = \bar{a}^i{}_j w^j$.

Solution to Problem 3.2

We are given that

$$v_i w^i = \bar{v}_i \bar{w}^i$$

for an *arbitrary* vector v_i, which transforms as $\bar{v}_i = \bar{a}_i{}^j v_j$. Substitute this relation in the above, and re-label indices on the r.h.s.

$$v_i w^i = \bar{a}_j{}^i v_i \bar{w}^j$$

Since v_i is arbitrary, we can equate coefficients

$$w^i = \bar{a}_j{}^i \bar{w}^j = a^i{}_j \bar{w}^j$$

where the second equality follows from Eqs. (3.36). Now invert this relation using Eqs. (3.32)

$$\bar{a}^i{}_j a^j{}_k = \delta^i{}_k$$

Hence, with a re-labeling of indices

$$\bar{w}^i = \bar{a}^i{}_j w^j$$

This is the desired result.

Problem 3.3 Choose cartesian coordinates $(q^1, q^2, q^3) = (x, y, z)$ in three-dimensional euclidian space and write the line element as

$$d\mathbf{s} = \hat{\mathbf{e}}_x \, dx + \hat{\mathbf{e}}_y \, dy + \hat{\mathbf{e}}_z \, dz$$

Here $(\hat{\mathbf{e}}_x, \hat{\mathbf{e}}_y, \hat{\mathbf{e}}_z)$ are the set of global, orthonormal, cartesian unit vectors.

(a) Show the metric is

$$\underline{g} = \begin{pmatrix} g_{xx} & g_{xy} & g_{xz} \\ g_{yx} & g_{yy} & g_{yz} \\ g_{zx} & g_{zy} & g_{zz} \end{pmatrix} = \begin{pmatrix} 1 & 0 & 0 \\ 0 & 1 & 0 \\ 0 & 0 & 1 \end{pmatrix}$$

(b) Show the reciprocal basis is identical to the original basis in this case, and hence there is no need to distinguish upper and lower indices if one sticks to cartesian coordinates in euclidian space.

Solution to Problem 3.3

(a) With the familiar set of global, orthonormal, cartesian unit vectors $(\hat{e}_x, \hat{e}_y, \hat{e}_z)$, one has

$$\hat{e}_i \cdot \hat{e}_j = \delta_{ij} \qquad ; \; (i,j) = (x.y.z)$$

where δ_{ij} is the usual Kronecker delta

$$\delta_{ij} = 1 \qquad ; \; \text{if } i = j$$
$$= 0 \qquad \text{if } i \neq j$$

Hence the metric is

$$g_{ij} = \hat{e}_i \cdot \hat{e}_j$$

$$\underline{g} = \begin{pmatrix} g_{xx} \; g_{xy} \; g_{xz} \\ g_{yx} \; g_{yy} \; g_{yz} \\ g_{zx} \; g_{zy} \; g_{zz} \end{pmatrix} = \begin{pmatrix} 1 \; 0 \; 0 \\ 0 \; 1 \; 0 \\ 0 \; 0 \; 1 \end{pmatrix}$$

where the last line is in matrix notation.

(b) The reciprocal basis is defined through Eq. (2.13)

$$\hat{e}^i \cdot \hat{e}_j = \delta^i{}_j$$
$$= 1 \qquad ; \; \text{if } i = j$$
$$= 0 \qquad \text{if } i \neq j$$

Hence in this case the reciprocal basis is *identical* to the original basis. Thus there is no need to distinguish upper and lower indices if one sticks to cartesian coordinates in euclidian space.

Problem 3.4 The point transformation from cartesian coordinates $(q^1, q^2, q^3) = (x, y, z)$ to spherical coordinates $(\xi^1, \xi^2, \xi^3) = (r, \theta, \phi)$ is given by

$$x = r \sin \theta \cos \phi \qquad ; \; y = r \sin \theta \sin \phi \qquad ; \; z = r \cos \theta$$

(a) Determine the transformation coefficients $a^i{}_j = \bar{a}_j{}^i$.

(b) Use the results in (a) to relate the components of a vector in spherical coordinates $\bar{v}_i$ to those in cartesian coordinates v_i.

(c) The metric in spherical coordinates is given in Eq. (5.178), and the metric in cartesian coordinates is given in Prob. 3.3. Determine the transformation coefficients $a_i{}^j = \bar{a}^j{}_i$ from the results found in (a).

Solution to Problem 3.4

(a) Define the following sets of coordinates in three-dimensions

$$q^i = (x, y, z)$$
$$\xi^i = (r, \theta, \phi)$$

The analytic transformation from cartesian to spherical coordinates is then

$$x = r \sin\theta \cos\phi \qquad ; \; y = r \sin\theta \sin\phi \qquad ; \; z = r \cos\theta$$

The transformation matrix in Eq. (3.31) is then given by

$$a^i{}_j = \frac{\partial q^i}{\partial \xi^j} = \begin{bmatrix} \sin\theta\cos\phi & r\cos\theta\cos\phi & -r\sin\theta\sin\phi \\ \sin\theta\sin\phi & r\cos\theta\sin\phi & r\sin\theta\cos\phi \\ \cos\theta & -r\sin\theta & 0 \end{bmatrix}$$

Note from Eq. (3.35)

$$a^i{}_j = (\mathbf{e}^i \cdot \boldsymbol{\alpha}_j) = \bar{a}_j{}^i$$

(b) The covariant components of a vector in spherical coordinates are then given by Eq. (3.49)

$$\bar{v}_j = \bar{a}_j{}^i v_i = a^i{}_j v_i$$

$$\begin{bmatrix} \bar{v}_r \\ \bar{v}_\theta \\ \bar{v}_\phi \end{bmatrix} = \begin{bmatrix} \sin\theta\cos\phi & \sin\theta\sin\phi & \cos\theta \\ r\cos\theta\cos\phi & r\cos\theta\sin\phi & -r\sin\theta \\ -r\sin\theta\sin\phi & r\sin\theta\cos\phi & 0 \end{bmatrix} \begin{bmatrix} v_x \\ v_y \\ v_z \end{bmatrix}$$

where the last line implies matrix multiplication.[1]

(c) The indices are raised and lowered according to Eqs. (3.13)

$$\mathbf{e}_i = g_{ij}\, \mathbf{e}^j$$
$$\boldsymbol{\alpha}^i = \bar{g}^{ij}\, \boldsymbol{\alpha}_j$$

Here, from Prob. 3.3 and Eq. (5.178)

$$g_{ij} = \begin{pmatrix} g_{xx} & g_{xy} & g_{xz} \\ g_{yx} & g_{yy} & g_{yz} \\ g_{zx} & g_{zy} & g_{zz} \end{pmatrix} = \begin{pmatrix} 1 & 0 & 0 \\ 0 & 1 & 0 \\ 0 & 0 & 1 \end{pmatrix}$$

$$\bar{g}_{ij} = \begin{bmatrix} g_{rr} & g_{r\theta} & g_{r\phi} \\ g_{\theta r} & g_{\theta\theta} & g_{\theta\phi} \\ g_{\phi r} & g_{\phi\theta} & g_{\phi\phi} \end{bmatrix} = \begin{bmatrix} 1 & 0 & 0 \\ 0 & r^2 & 0 \\ 0 & 0 & r^2\sin^2\theta \end{bmatrix}$$

---

[1]Note $a^i{}_j \equiv (\underline{a})_{ij} = (\underline{a}^T)_{ji}$.

and from Eqs. (2.21)

$$\bar{g}^{ij} = \left(\underline{\bar{g}}^{-1}\right)_{ij} = \begin{bmatrix} 1 & 0 & 0 \\ 0 & 1/r^2 & 0 \\ 0 & 0 & 1/r^2 \sin^2\theta \end{bmatrix} = \bar{g}^{ji}$$

It follows that

$$a_i{}^j = (\mathbf{e}_i \cdot \boldsymbol{\alpha}^j) = g_{ii'}\,\bar{g}^{jj'}(\mathbf{e}^{i'} \cdot \boldsymbol{\alpha}_{j'}) = g_{ii'}\,\bar{g}^{jj'}\,a^{i'}{}_{j'} = \bar{a}^j{}_i$$

In matrix notation, this is

$$\underline{g}\,\underline{a}\,\underline{\bar{g}}^{-1} = \underline{\bar{a}}^T$$

Thus $\underline{\bar{a}}^T$ is given by the following matrix product

$$\underline{\bar{a}}^T = \begin{bmatrix} 1 & 0 & 0 \\ 0 & 1 & 0 \\ 0 & 0 & 1 \end{bmatrix} \times$$

$$\begin{bmatrix} \sin\theta\cos\phi & r\cos\theta\cos\phi & -r\sin\theta\sin\phi \\ \sin\theta\sin\phi & r\cos\theta\sin\phi & r\sin\theta\cos\phi \\ \cos\theta & -r\sin\theta & 0 \end{bmatrix} \begin{bmatrix} 1 & 0 & 0 \\ 0 & 1/r^2 & 0 \\ 0 & 0 & 1/r^2\sin^2\theta \end{bmatrix}$$

$$= \begin{bmatrix} \sin\theta\cos\phi & \cos\theta\cos\phi/r & -\sin\phi/r\sin\theta \\ \sin\theta\sin\phi & \cos\theta\sin\phi/r & \cos\phi/r\sin\theta \\ \cos\theta & -\sin\theta/r & 0 \end{bmatrix}$$

The equality $a_i{}^j = \bar{a}^j{}_i = (\underline{\bar{a}}^T)_{ij}$ follows from Eqs. (3.36).
Let us check to see if Eq. (3.32) is satisfied

$$\underline{a}\,\underline{\bar{a}} = \underline{1}$$

We have

$$\begin{bmatrix} \sin\theta\cos\phi & r\cos\theta\cos\phi & -r\sin\theta\sin\phi \\ \sin\theta\sin\phi & r\cos\theta\sin\phi & r\sin\theta\cos\phi \\ \cos\theta & -r\sin\theta & 0 \end{bmatrix} \begin{bmatrix} \sin\theta\cos\phi & \sin\theta\sin\phi & \cos\theta \\ \cos\theta\cos\phi/r & \cos\theta\sin\phi/r & -\sin\theta/r \\ -\sin\phi/r\sin\theta & \cos\phi/r\sin\theta & 0 \end{bmatrix}$$

$$= \begin{bmatrix} 1 & 0 & 0 \\ 0 & 1 & 0 \\ 0 & 0 & 1 \end{bmatrix}$$

and, indeed, it checks!

Problem 3.5 Another example of a tensor is the stress tensor for non-viscous fluids, whose components in cartesian coordinates are given by

$$T_{ij} = p\,g_{ij} + \rho\,v_i v_j$$

Here p is the pressure and ρ is the mass density, both quantities being defined in the rest frame of the fluid. The vector $\mathbf{v}$ is the fluid velocity, and g_{ij} is the metric.[2]

(a) Prove that T_{ij} transforms as a second-rank tensor.

(b) Use the results in Prob. 3.4 to determine the components of the stress tensor in spherical coordinates.

Solution to Problem 3.5

(a) For simplicity, first raise the indices and write

$$T^{ij} = p\,g^{ij} + \rho\,v^i v^j$$

It is proven in Eqs. (3.51)–(3.56) that g^{ij} is a second-rank tensor

$$g^{ij} = a^i{}_{i'} a^j{}_{j'} \bar{g}^{i'j'}$$

Furthermore, from Eq. (3.47), v^i is a first-rank tensor (vector)

$$v^i = a^i{}_j \bar{v}^j$$

Since both p and ρ are scalars, defined in the rest frame of the fluid, one has

$$T^{ij} = a^i{}_{i'} a^j{}_{j'} \bar{T}^{i'j'}$$

Thus by Eqs. (3.50), T^{ij} transforms as a second-rank tensor.

(b) From Eqs. (3.50), the stress tensor in spherical coordinates is given by

$$\bar{T}^{ij} = \bar{a}^i{}_{i'} \bar{a}^j{}_{j'} T^{i'j'}$$

where, from Prob. 3.4,

$$\bar{a}^i{}_j = (\underline{\bar{a}})_{ij}$$

---

[2]See [Fetter and Walecka (2003), p. 297]. Note that with cartesian coordinates $g_{ij} = \delta_{ij}$ is the usual Kronecker delta. We use the terms *stress tensor* and *energy-momentum tensor* interchangeably for such a fluid in this book.

with

$$\underline{\bar{a}} = \begin{bmatrix} \sin\theta\cos\phi & \sin\theta\sin\phi & \cos\theta \\ \cos\theta\cos\phi/r & \cos\theta\sin\phi/r & -\sin\theta/r \\ -\sin\phi/r\sin\theta & \cos\phi/r\sin\theta & 0 \end{bmatrix}$$

Problem 3.6 There are eight elements in the affine connection Γ^i_{jk} in plane polar coordinates. They are given in Eqs. (3.84). Five are computed in the text.

(a) Verify the remaining three.

(b) The change in basis vectors as one moves to a neighboring point $(r+dr, \phi+d\phi)$ is given in terms of the affine connection by Eqs. (3.85). Provide a direct geometrical derivation of these results.

Solution to Problem 3.6

(a) The metric and its inverse in plane polar coordinates are given in Eqs. (3.83)

$$\underline{g} = \begin{pmatrix} g_{rr} & g_{r\phi} \\ g_{\phi r} & g_{\phi\phi} \end{pmatrix} = \begin{pmatrix} 1 & 0 \\ 0 & r^2 \end{pmatrix}$$

$$\underline{g}^{-1} = \begin{pmatrix} g^{rr} & g^{r\phi} \\ g^{\phi r} & g^{\phi\phi} \end{pmatrix} = \begin{pmatrix} 1 & 0 \\ 0 & 1/r^2 \end{pmatrix}$$

Both are diagonal. The affine connection is given by Eqs. (3.84)

$$\Gamma^\phi_{r\phi} = \Gamma^\phi_{\phi r} = \frac{1}{r}$$

$$\Gamma^r_{\phi\phi} = -r \qquad ; \text{ all others vanish}$$

The three relations that are not derived in the text are the following

$$\Gamma^\phi_{r\phi} = \Gamma^\phi_{\phi r} = \frac{1}{2}g^{\phi\phi}\left[\frac{\partial g_{r\phi}}{\partial\phi} + \frac{\partial g_{\phi\phi}}{\partial r} - \frac{\partial g_{r\phi}}{\partial\phi}\right] = \frac{1}{2}\left(\frac{1}{r^2}\right)(2r) = \frac{1}{r}$$

$$\Gamma^\phi_{rr} = \frac{1}{2}g^{\phi\phi}\left[\frac{\partial g_{r\phi}}{\partial r} + \frac{\partial g_{r\phi}}{\partial r} - \frac{\partial g_{rr}}{\partial\phi}\right] = 0$$

(b) The change in basis vectors as one moves to a neighboring point $(r+dr, \phi+d\phi)$ is given in terms of the affine connection by Eqs. (3.85)

$$d\,\mathbf{e}_r = \frac{1}{r}d\phi\,\mathbf{e}_\phi$$

$$d\,\mathbf{e}_\phi = \frac{1}{r}dr\,\mathbf{e}_\phi - rd\phi\,\mathbf{e}_r$$

Consider Fig. 2.5 in the text. If the point at (r, ϕ) is moved to $(r + dr, \phi)$, then $\mathbf{e}_r = \hat{\mathbf{e}}_r$ does not change, and from Eqs. (2.26), $\mathbf{e}_\phi$ changes by

$$d\mathbf{e}_\phi = dr\,\hat{\mathbf{e}}_\phi = \frac{1}{r}dr\,\mathbf{e}_\phi$$

If the point is moved to $(r, \phi+d\phi)$, then both basis vectors change according to

$$d\hat{\mathbf{e}}_r = d\phi\,\hat{\mathbf{e}}_\phi = \frac{1}{r}d\phi\,\mathbf{e}_\phi \qquad ; \quad \frac{1}{r}d\mathbf{e}_\phi = d\hat{\mathbf{e}}_\phi = -d\phi\,\hat{\mathbf{e}}_r$$

Problem 3.7 Write the gradient as $\boldsymbol{\nabla} = \mathbf{e}_i\,\nabla^i = \mathbf{e}^i\,\nabla_i$ and show that

$$\nabla^i = g^{ij}\,\nabla_j$$

Solution to Problem 3.7

Use the fact that the metric raises the index according to

$$\mathbf{e}^i = g^{ij}\,\mathbf{e}_j$$

Hence with a re-labeling of indices, and the observation that the metric is symmetric, one has

$$\mathbf{e}^i\,\nabla_i = g^{ij}\,\mathbf{e}_j\nabla_i = \mathbf{e}_i\,g^{ij}\nabla_j = \mathbf{e}_i\,\nabla^i$$

Hence

$$\nabla^i = g^{ij}\,\nabla_j$$

Problem 3.8 Let $\hat{\mathbf{e}}_k$ with $k = (1, \cdots, n)$ be a global, complete, orthonormal set of cartesian unit vectors in the n-dimensional euclidian space.

(a) Show that any vector field $\mathbf{v}(q^1, \cdots, q^n)$ can be expanded in this basis according to

$$\mathbf{v}(q^1, \cdots, q^n) = v^k(q^1, \cdots, q^n)\,\hat{\mathbf{e}}_k$$

(b) Hence conclude that the rule for interchanging the order of differentiation when the second partial derivative $\partial^2/\partial q^i\partial q^j$ is applied to this vector field is just that for multivariable functions. What is the rule?

Solution to Problem 3.8

(a) Let $\hat{\mathbf{e}}_k$ with $k = (1, \cdots, n)$ be a global, complete, orthonormal set of cartesian unit vectors in the n-dimensional euclidian space. At each point in the space, an arbitrary vector field $\mathbf{v}(q^1, \cdots, q^n)$ can be expanded in this basis according to

$$\mathbf{v}(q^1, \cdots, q^n) = v^k(q^1, \cdots, q^n)\,\hat{\mathbf{e}}_k$$

(b) Since the basis vectors are global, they do not depend on position. Hence the second derivative of the vector field is given by[3]

$$\frac{\partial^2 \mathbf{v}}{\partial q^i \partial q^j} = \frac{\partial^2 v^k}{\partial q^i \partial q^j}\,\hat{\mathbf{e}}_k$$

Since partial derivatives of ordinary multi-variable functions commute, one has

$$\frac{\partial^2 v^k}{\partial q^i \partial q^j} = \frac{\partial^2 v^k}{\partial q^j \partial q^i}$$

This is to be contrasted with the second covariant derivative in a riemannian space in Eq. (5.60).

---

[3] Compare the second covariant derivative in Eq. (5.57).

Chapter 4

Particle on a Two-Dimensional Surface–Revisited

Problem 4.1 (a) Show from Eq. (4.8) that when the particle motion is constrained to the surface

$$\left(\frac{d\mathbf{v}}{dt}\right)_{\perp} = v^i v^j \Gamma^3_{ij} \mathbf{e}_3 \qquad ; (i,j) = (1,2)$$

(b) Find the constraint force then implied by Eq. (4.9).

(c) Discuss the implications for Γ^3_{ij} for the curved surface.

Solution to Problem 4.1

(a) Figure 4.1 in the text illustrates the situation where a point particle is constrained to move without friction on a two-dimensional surface of arbitrary shape. The generalized coordinates (q^1, q^2, q^3) locate the particle in three-dimensional euclidian space. The basis vector $\mathbf{e}_3$ is normal to the tangent plane, and the velocity $\mathbf{v}$ lies in it. The acceleration is given in Eq. (4.8)

$$\frac{d\mathbf{v}}{dt} = \left(\frac{dv^k}{dt} + v^i v^j \Gamma^k_{ij}\right) \mathbf{e}_k \qquad ; (i,j,k) = (1,2,3)$$

Since the particle is confined to the surface, one has $\mathbf{v} = (v^1, v^2, 0) = (dq^1/dt, dq^2/dt, 0)$, and hence the third component of the above relation gives the acceleration normal to the surface as

$$\left(\frac{d\mathbf{v}}{dt}\right)_{\perp} = v^i v^j \Gamma^3_{ij} \mathbf{e}_3 \qquad ; (i,j) = (1,2)$$

(b) Newton's second law gives Eq. (4.9)

$$m\frac{d\mathbf{v}}{dt} = \mathbf{F}_{\perp} = F^3 \mathbf{e}_3$$

15

The third component of this relation then yields the constraint force

$$\mathbf{F}_\perp = m\, v^i v^j \Gamma^3_{ij}\, \mathbf{e}_3 \qquad\quad ; \ (i,j) = (1,2)$$

(c) Since a curved surface exerts a non-zero constraint force to keep the particle confined to it, we conclude that Γ^3_{ij} must be non-zero when this happens. Indeed, from Eq. (4.6)

$$d\mathbf{e}_i = \Gamma^k_{ij} dq^j\, \mathbf{e}_k$$

and we observe that Γ^3_{ij} will be non-zero whenever the vectors in the tangent plane acquire some third component upon a change in position.

Problem 4.2 (a) Start from Eq. (4.22). Write out in detail the three steps given below that equation and show that one indeed arrives at Lagrange's Eqs. (4.12).

(b) Provide all the steps for the derivation of Lagrange's Eqs. (4.12) on the surface from the condition of parallel displacement in the tangent plane given in Eq. (4.25).

(c) Provide all the steps for the derivation of the geodesic Eqs. (4.27) on the surface from the condition of parallel displacement in the tangent plane stated there.

Solution to Problem 4.2

(a) In the tangent plane, we have at the point (q^1, q^2) [see Eq. (4.22)][1]

$$\mathbf{v} = v^1\, \mathbf{e}_1 + v^2\, \mathbf{e}_2$$

From Eq. (4.7), the change $d\mathbf{v}_\parallel$ in the tangent plane is

$$d\mathbf{v}_\parallel = \left(dv^k + \Gamma^k_{ij} dq^j v^i\right) \mathbf{e}_k \qquad\quad ; \ (i,j,k) = (1,2)$$

Here, in employing Eq. (4.6), we have used the fact that $dq^3 = 0$ on the surface.

Parallel displacement in the tangent plane implies that $d\mathbf{v}_\parallel = 0$ [see Eq. (4.21)] or, equivalently,

$$\frac{d\mathbf{v}_\parallel}{dt} = 0 \qquad\qquad ; \ \text{parallel displacement}$$

---

[1]Note that $v^3 = dq^3/dt = 0$ everywhere on the surface.

Since $v^k = dq^k/dt$, and $\Gamma_{ij}^k = \Gamma_{ji}^k$, this condition becomes

$$\left(\frac{d^2 q^k}{dt^2} + \Gamma_{ij}^k \frac{dq^i}{dt}\frac{dq^j}{dt}\right)\mathbf{e}_k = 0 \qquad ; (i,j,k) = (1,2)$$

Linear independence of the basis vectors $(\mathbf{e}_1, \mathbf{e}_2)$ then yields Lagrange's Eqs. (4.12)

$$\frac{d^2 q^k}{dt^2} + \Gamma_{ij}^k \frac{dq^i}{dt}\frac{dq^j}{dt} = 0 \qquad ; (i,j,k) = (1,2)$$

(b) The proof here is just that in part (a). Note that all quantities are now confined to the tangent plane.

(c) Identify the unit vector $\mathbf{u}$ tangent to the trajectory C on the surface, and distance s along the curve, according to Eq. (4.26)

$$\mathbf{u} = \frac{1}{v_0}\mathbf{v}$$

$$t = \frac{1}{v_0}s$$

Since energy is conserved, the magnitude of the velocity v_0 is constant along the trajectory. The condition of parallel displacement in the tangent plane in Eqs. (4.27) then reduces to

$$d\mathbf{u} = d\mathbf{v}_{\|} = 0$$

From part (a), this is

$$\left(dv^k + \Gamma_{ij}^k dq^j v^i\right)\mathbf{e}_k = 0 \qquad ; (i,j,k) = (1,2)$$

Write $v^i = v_0\, dq^i/ds$, and divide by $v_0\, ds$, to arrive at

$$\left(\frac{d^2 q^k}{ds^2} + \Gamma_{ij}^k \frac{dq^i}{ds}\frac{dq^j}{ds}\right)\mathbf{e}_k = 0 \qquad ; (i,j,k) = (1,2)$$

Linear independence of the basis vectors $(\mathbf{e}_1, \mathbf{e}_2)$ then yields the geodesic Eqs. (4.27)

$$\frac{d^2 q^k}{ds^2} + \Gamma_{ij}^k \frac{dq^i}{ds}\frac{dq^j}{ds} = 0 \qquad ; (i,j,k) = (1,2)$$

Once again, all quantities are confined to the tangent plane.

Chapter 5

Some Tensor Analysis

Problem 5.1 It is shown in the text that the covariant derivative of a vector is given by

$$v^i{}_{;j} = \frac{\partial v^i}{\partial q^j} + \Gamma^i_{jk} v^k$$

$$v_{i;j} = \frac{\partial v_i}{\partial q^j} - \Gamma^k_{ij} v_k$$

It is also shown that the metric is constant under covariant differentiation so that it can be moved through the derivative to raise and lower indices. Thus

$$g_{ik} \left(v^k{}_{;j} \right) = v_{i;j}$$

Use this relation to derive the second of the above equations from the first.

Solution to Problem 5.1

The *covariant derivative* of the vector field is identified in Eq. (5.3)

$$v^k{}_{;j} \equiv \frac{\partial v^k}{\partial q^j} + v^i \Gamma^k_{ij}$$

The metric tensor is given by

$$\underline{\underline{g}} = g_{ij} \mathbf{e}^i \mathbf{e}^j$$

The differentials of the basis vectors satisfy Eqs. (5.24)

$$d\,\mathbf{e}_i = \Gamma^k_{ij}\, \mathbf{e}_k dq^j \qquad ;\; d\mathbf{e}^i = -\Gamma^i_{kj}\, \mathbf{e}^k dq^j$$

19

As in Eqs. (5.73)–(5.75), the differential of the metric tensor is then

$$d\underline{\underline{g}} = \left(dg_{ij} - \Gamma_{il}^m\, g_{mj}dq^l - \Gamma_{jl}^m\, g_{mi}dq^l\right) \mathbf{e}^i \mathbf{e}^j$$

The differential of g_{ij} is

$$dg_{ij} = d\left(\mathbf{e}_i \cdot \mathbf{e}_j\right) = d\mathbf{e}_i \cdot \mathbf{e}_j + \mathbf{e}_i \cdot d\mathbf{e}_j$$
$$= \Gamma_{il}^m\, g_{mj}dq^l + \Gamma_{jl}^m\, g_{mi}dq^l$$

Hence, exactly as in Eqs. (5.78)–(5.80), this covariant derivative of the metric tensor also vanishes

$$d\underline{\underline{g}} = \left(\mathcal{D}g_{ij}\right)\mathbf{e}^i\mathbf{e}^j = 0$$

$$g_{ij;\,k} = \frac{\mathcal{D}g_{ij}}{\partial q^k} = 0$$

Covariant differentiation is distributive. Therefore

$$v_{i;j} = \left(g_{il}v^l\right)_{;j} = g_{il;j}\, v^l + g_{il}\, v^l_{\;;j}$$
$$= g_{ik}\left(v^k_{\;;j}\right)$$

Thus, although not at all immediately obvious from the expressions that appear on the r.h.s., the *metric allows us to lower (and raise) the index on the vector which is undergoing covariant differentiation.*

Let's see how this works in detail. Consider

$$g_{ik}\left(v^k_{\;;j}\right) = g_{ik}\frac{\partial v^k}{\partial q^j} + g_{ik}\Gamma_{jl}^k v^l$$
$$= \frac{\partial(g_{ik}v^k)}{\partial q^j} - v^k\frac{\partial g_{ik}}{\partial q^j} + g_{ik}\Gamma_{jl}^k v^l$$

Now use the above result for $\partial g_{ik}/\partial q^j$ in the second term, and change dummies in the last term

$$g_{ik}\left(v^k_{\;;j}\right) = \frac{\partial v_i}{\partial q^j} - v^k\left[\Gamma_{ij}^m g_{mk} + \Gamma_{kj}^m g_{mi}\right] + g_{im}\Gamma_{jk}^m v^k$$
$$= \frac{\partial v_i}{\partial q^j} - \Gamma_{ij}^m v_m = v_{i;j}$$

Problem 5.2 Prove that the symmetry of the Ricci tensor $R_{ij} = R_{ji}$ follows directly from the symmetry property of the Riemann tensor $R_{jklm} = R_{lmjk}$ derived in Eq. (5.120).

Solution to Problem 5.2

The Ricci tensor is defined in terms of the Riemann tensor in Eq. (5.104)

$$R_{ij} \equiv R^k{}_{ikj} \qquad ; \text{ Ricci tensor}$$

Write out the contraction

$$R^k{}_{ikj} = g^{kl} R_{likj}$$

Now use the symmetry property of the Riemann tensor in Eq. (5.120)

$$R_{likj} = R_{kjli}$$

This gives

$$R^k{}_{ikj} = g^{kl} R_{kjli} = R^l{}_{jli}$$

This establishes the symmetry of the Ricci tensor

$$R_{ij} = R_{ji}$$

Problem 5.3 Show the symmetry properties of the Riemann tensor are preserved under a coordinate transformation.

Solution to Problem 5.3

Under a coordinate transformation, the Riemann tensor transforms according to

$$\bar{R}_{ijkl} = \bar{a}_i{}^{i'} \bar{a}_j{}^{j'} \bar{a}_k{}^{k'} \bar{a}_l{}^{l'} R_{i'j'k'l'}$$

Consider any interchange of the indices on $R_{i'j'k'l'}$. This is immediately converted to a corresponding interchange of the indices on $\bar{R}_{ijkl}$ through a re-shuffling of the transformation matrices. For example, introduce the symmetry property

$$R_{i'j'k'l'} = R_{k'l'i'j'}$$

Then

$$\bar{R}_{ijkl} = \bar{a}_i{}^{i'} \bar{a}_j{}^{j'} \bar{a}_k{}^{k'} \bar{a}_l{}^{l'} R_{k'l'i'j'} = \bar{a}_k{}^{k'} \bar{a}_l{}^{l'} \bar{a}_i{}^{i'} \bar{a}_j{}^{j'} R_{k'l'i'j'} = \bar{R}_{klij}$$

and the symmetry is immediately reflected in the transformed $\bar{R}_{ijkl}$.

Problem 5.4[1] Choose cartesian coordinates (x, y, z) in three-dimensional euclidian space as in Prob. 3.3.
    (a) What is the affine connection?
    (b) What is the Riemann tensor?
    (c) What are the Ricci tensor and scalar curvature?

Solution to Problem 5.4

The metric in cartesian coordinates in three-dimensional euclidian space is given in Prob. 3.3

$$\begin{pmatrix} g_{xx} & g_{xy} & g_{xz} \\ g_{yx} & g_{yy} & g_{yz} \\ g_{zx} & g_{zy} & g_{zz} \end{pmatrix} = \begin{pmatrix} g^{xx} & g^{xy} & g^{xz} \\ g^{yx} & g^{yy} & g^{yz} \\ g^{zx} & g^{zy} & g^{zz} \end{pmatrix} = \begin{pmatrix} 1 & 0 & 0 \\ 0 & 1 & 0 \\ 0 & 0 & 1 \end{pmatrix}$$

(a) From Eq. (3.79), the affine connection is

$$\Gamma^k_{ij} = \frac{1}{2} g^{km} \left[\frac{\partial g_{mi}}{\partial q^j} + \frac{\partial g_{mj}}{\partial q^i} - \frac{\partial g_{ij}}{\partial q^m} \right] \qquad ; (i, j, k) = 1, 2, 3$$

Since the metric is global, with no coordinate dependence, this vanishes

$$\Gamma^k_{ij} = 0$$

(b) The Riemann curvature tensor in this space is given in Eq. (5.47)

$$R^i_{\ jlk} = \left(\frac{\partial}{\partial q^l} \Gamma^i_{jk} + \Gamma^i_{lm} \Gamma^m_{jk} \right) - (k \leftrightarrow l) \qquad ; (i, j, k, l) = 1, 2, 3$$

$$= \left(\frac{\partial}{\partial q^l} \Gamma^i_{jk} + \Gamma^i_{lm} \Gamma^m_{jk} \right) - \left(\frac{\partial}{\partial q^k} \Gamma^i_{jl} + \Gamma^i_{km} \Gamma^m_{jl} \right)$$

Since the affine connection vanishes, this does also

$$R^i_{\ jlk} = 0$$

(c) The Ricci tensor and scalar curvature are obtained from the Riemann tensor through Eqs. (5.104)–(5.105)

$$R_{ij} \equiv R^k_{\ ikj} \qquad ; \text{Ricci tensor}$$

$$R \equiv R^i_{\ i} = R^{ki}_{\ \ ki} \qquad ; \text{scalar curvature}$$

These again vanish

$$R_{ij} = R = 0$$

We say the space is "flat".

---

[1]Problems 5.4 and 5.5 are very easy, but they are also very instructive.

Problem 5.5 Generalize the results in Prob. 5.4 to n-dimensional euclidian space. Show the Riemann tensor, Ricci tensor, and scalar curvature all vanish.

Solution to Problem 5.5

The results in Prob. 5.4 are generalized to n-dimensional euclidian space by just letting the indices run from 1 to n

$$(i, j, k, l) = 1, 2, \cdots, n$$

The metric is the n-dimensional unit matrix. The Riemann tensor, Ricci tensor, and scalar curvature again all vanish

$$R_{ijlk} = 0$$
$$R_{ij} = 0$$
$$R = 0$$

Problem 5.6 The tensor transformation law in Eq. (3.50) not only provides insight but can also save an incredible amount of work. Start from the cartesian coordinates in n-dimensional euclidian space of Prob. 5.5, and make a point transformation to *any new set of generalized coordinates*.

(a) What is the Riemann tensor in the new coordinate system?

(b) What are the Ricci tensor and scalar curvature in the new coordinate system?

Solution to Problem 5.6

(a) The tensor transformation law from the cartesian coordinates in n-dimensional euclidian space in Prob. 5.5 to *any* new set of generalized coordinates is [see Eqs. (3.50)]

$$\bar{R}_{ijkl} = \bar{a}_i^{\ i'} \bar{a}_j^{\ j'} \bar{a}_k^{\ k'} \bar{a}_l^{\ l'} R_{i'j'k'l'}$$

No matter how difficult the direct calculation of $\bar{R}_{ijkl}$ (see the expressions in Prob. 5.4), the result will again clearly *vanish*

$$\bar{R}_{ijkl} = 0$$

(b) Since both the Ricci tensor and scalar curvature are obtained from $\bar{R}_{ijkl}$, they also vanish

$$\bar{R}_{ij} = \bar{R} = 0$$

Problem 5.7 Consider a flat, two-dimensional euclidian space and the point transformation from cartesian coordinates (x, y) to plane polar coordinates (r, ϕ) (see Fig. 2.5 in the text)

$$x = r \cos \phi \qquad ; \ y = r \sin \phi$$

The affine connection in cartesian coordinates vanishes by the argument in Prob. 5.4(a), while the affine connection in plane polar coordinates is given in Eqs. (3.84). Use these specific results to conclude that the affine connection Γ^i_{jk} does *not* transform as a third-rank tensor.

Solution to Problem 5.7

As in Prob. 3.4(a), the transformation matrix from cartesian coordinates in two-dimensions to plane polar coordinates is

$$a^i_{\ j} = \frac{\partial q^i}{\partial \xi^j} = \begin{pmatrix} \cos \phi & -r \sin \phi \\ \sin \phi & r \cos \phi \end{pmatrix} = \bar{a}_j^{\ i}$$

The matrices $a_i^{\ j} = \bar{a}^j_{\ i}$ are obtained from this as in Prob. 3.4(c).

In cartesian coordinates, the affine connection vanishes (see Prob. 5.4)

$$\Gamma^k_{ij} = 0$$

If Γ^k_{ij} were a third-rank tensor, then the result in plane polar coordinates would be given by

$$\bar{\Gamma}^k_{ij} = \bar{a}^k_{\ k'} \, \bar{a}_i^{\ i'} \bar{a}_j^{\ j'} \Gamma^{k'}_{i'j'}$$

This clearly vanishes[2]

$$\bar{\Gamma}^k_{ij} = 0 \qquad ; \ \underline{\text{not}} \text{ true}$$

Direct calculation of $\bar{\Gamma}^k_{ij}$ in plane polar coordinates in Eqs. (3.84) indicates this is <u>not</u> the case. *Hence we conclude that Γ^k_{ij} cannot be a third-rank tensor.*

Problem 5.8 Cylindrical coordinates are a three-dimensional set (r, z, ϕ) where z is the height above the plane in Fig. 2.5 in the text. The line element in cylindrical coordinates in three-dimensional euclidian space can be written

$$d\mathbf{s} = \hat{\mathbf{e}}_r \, dr + \hat{\mathbf{e}}_z \, dz + \hat{\mathbf{e}}_\phi \, r d\phi$$

---

[2]See Prob. 5.6.

Here $(\hat{e}_r, \hat{e}_z, \hat{e}_\phi)$ form a set of orthonormal unit vectors. The surface of a right circular cylinder is then described with the coordinates (z, ϕ) and the condition $r = $ constant.

(a) Show the metric on the surface of a cylinder with radius r is

$$\underline{g} = \begin{pmatrix} g_{zz} & g_{z\phi} \\ g_{\phi z} & g_{\phi\phi} \end{pmatrix} = \begin{pmatrix} 1 & 0 \\ 0 & r^2 \end{pmatrix}$$

(b) Calculate the affine connection.

(c) Calculate the Riemann tensor, Ricci tensor, and scalar curvature.

(d) Could you have anticipated these results by considering the parallel transport of a vector around a closed loop with two sides along the axis of the cylinder and two sides along the circular circumference?

Solution to Problem 5.8

(a) The radius r is constant on the surface of the cylinder, and the coordinates are $q^i = (z, \phi)$. The square of the line element is then

$$(ds)^2 = (dz)^2 + r^2 (d\phi)^2$$

Thus the metric, and its inverse, are

$$\begin{pmatrix} g_{zz} & g_{z\phi} \\ g_{\phi z} & g_{\phi\phi} \end{pmatrix} = \begin{pmatrix} 1 & 0 \\ 0 & r^2 \end{pmatrix} \qquad ; \begin{pmatrix} g^{zz} & g^{z\phi} \\ g^{\phi z} & g^{\phi\phi} \end{pmatrix} = \begin{pmatrix} 1 & 0 \\ 0 & 1/r^2 \end{pmatrix}$$

(b) Since r is constant, and the coordinates are (z, ϕ), the affine connection vanishes

$$\Gamma_{ij}^k = \frac{1}{2} g^{km} \left[\frac{\partial g_{mi}}{\partial q^j} + \frac{\partial g_{mj}}{\partial q^i} - \frac{\partial g_{ij}}{\partial q^m} \right] = 0 \qquad ; (i, j, k) = (z, \phi)$$

(c) The Riemann tensor then also vanishes

$$R^i_{jlk} = \left(\frac{\partial}{\partial q^l} \Gamma_{jk}^i + \Gamma_{lm}^i \Gamma_{jk}^m \right) - (k \leftrightarrow l) = 0 \qquad ; (i, j, k, l) = (z, \phi)$$

as do the Ricci tensor and scalar curvature

$$R_{ij} = R^k_{ikj} = 0$$
$$R = R^i_{\ i} = R^{ki}_{\ \ ki} = 0$$

The surface of the cylinder is *flat*.[3]

---

[3] Just roll it out — it does not *pucker*.

(d) Consider parallel transport of a vector **v**, as illustrated in Fig. 5.2 in the text, about the following closed loop on the surface of the cylinder: along an arc perpendicular to the axis of the cylinder; up a line in the surface parallel to the axis; back along an arc perpendicular to the axis; and back down a line parallel to the axis. The vector **v** does not rotate over this cycle. Hence, by Eq. (5.44)

$$\Delta v^i = -\frac{1}{2} R^i{}_{jlk} v_0^j S^{lk} = 0$$

This indicates that the Riemann tensor vanishes on the surface of the cylinder.[4]

Problem 5.9 The metric in cylindrical coordinates (r, z, ϕ) follows directly from the line element in Prob. 5.8

$$\underline{g} = \begin{pmatrix} g_{rr} & g_{rz} & g_{r\phi} \\ g_{zr} & g_{zz} & g_{z\phi} \\ g_{\phi r} & g_{\phi z} & g_{\phi\phi} \end{pmatrix} = \begin{pmatrix} 1 & 0 & 0 \\ 0 & 1 & 0 \\ 0 & 0 & r^2 \end{pmatrix}$$

(a) What is the covariant divergence in cylindrical coordinates?
(b) What is the volume element?

Solution to Problem 5.9

(a) In cylindrical coordinates, $q^i = (r, z, \phi)$. From Prob. 5.8, the line element and interval are

$$d\mathbf{s} = \hat{\mathbf{e}}_r \, dr + \hat{\mathbf{e}}_z \, dz + \hat{\mathbf{e}}_\phi r d\phi$$
$$(d\mathbf{s})^2 = (dr)^2 + (dz)^2 + (rd\phi)^2$$

The metric, and its inverse, are

$$\begin{pmatrix} g_{rr} & g_{rz} & g_{r\phi} \\ g_{zr} & g_{zz} & g_{z\phi} \\ g_{\phi r} & g_{\phi z} & g_{\phi\phi} \end{pmatrix} = \begin{pmatrix} 1 & 0 & 0 \\ 0 & 1 & 0 \\ 0 & 0 & r^2 \end{pmatrix} \quad ; \quad \begin{pmatrix} g^{rr} & g^{rz} & g^{r\phi} \\ g^{zr} & g^{zz} & g^{z\phi} \\ g^{\phi r} & g^{\phi z} & g^{\phi\phi} \end{pmatrix} = \begin{pmatrix} 1 & 0 & 0 \\ 0 & 1 & 0 \\ 0 & 0 & 1/r^2 \end{pmatrix}$$

The determinant of the metric $(\underline{g})_{ij} = g_{ij}$ is

$$g = \det \underline{g} = r^2$$

---

[4] At least all its relevant components vanish.

As in Eqs. (5.176), a vector field **v** can be expanded in cylindrical coordinates as

$$\mathbf{v} = v^r \hat{\mathbf{e}}_r + v^z \hat{\mathbf{e}}_z + (rv^\phi)\hat{\mathbf{e}}_\phi$$

The covariant divergence is given in Eq, (5.173)

$$v^l{}_{;l} = \frac{1}{\sqrt{g}} \frac{\partial}{\partial q^l} \left(\sqrt{g} \, v^l \right) \qquad ; l = (r, z, \phi)$$

Hence, as in Eqs. (5.180)–(5.181), the covariant divergence in cylindrical coordinates is

$$v^i{}_{;i} = \frac{1}{r} \left[\frac{\partial}{\partial r}(rv^r) + \frac{\partial}{\partial z}(rv^z) + \frac{\partial}{\partial \phi}(rv^\phi) \right]$$
$$= \frac{1}{r}\frac{\partial}{\partial r}(rv^r) + \frac{\partial}{\partial z}(v^z) + \frac{1}{r}\frac{\partial}{\partial \phi}(rv^\phi)$$

This is the same expression for the divergence in cylindrical coordinates that appears in [Fetter and Walecka (2003)]

$$v^i{}_{;i} = \boldsymbol{\nabla} \cdot \mathbf{v}$$

(b) The volume element in cylindrical coordinates is given by Eq. (5.151)

$$dV = \sqrt{g}\, dq^1 dq^2 dq^3 = r\, dr dz d\phi$$

Problem 5.10 Use polar coordinates in the $z = 0$ plane as generalized coordinates $(q^1, q^2) = (r, \phi)$. Consider the problem of a particle of mass m moving without friction, and with only the constraint force, on a surface of revolution $z = z(r)$ about the z-axis. Here $z(r)$ is assumed to be smooth and single-valued.

(a) Show the square of the line element on the surface is given by

$$(ds)^2 = \left\{ 1 + \left[\frac{dz(r)}{dr} \right]^2 \right\} (dr)^2 + (rd\phi)^2$$

(b) What is the lagrangian?
(c) What are Lagrange's equations?
(d) What is the metric?

(e) Sketch how the motion of the particle on the surface is reflected in the space of the generalized coordinates.[5]

Solution to Problem 5.10

(a) From Prob. 5.9, the line element and interval in cylindrical coordinates are

$$ds = \hat{e}_r \, dr + \hat{e}_z \, dz + \hat{e}_\phi r d\phi$$
$$(ds)^2 = (dr)^2 + (dz)^2 + (rd\phi)^2$$

For the surface of revolution

$$z = z(r)$$
$$dz = \frac{dz(r)}{dr} dr$$

Therefore, the line element on the surface is

$$(ds)^2 = \left[1 + \left(\frac{dz}{dr} \right)^2 \right] (dr)^2 + (rd\phi)^2$$

(b) The lagrangian for a particle of mass m moving on the surface is

$$L(r, \phi, \dot{r}, \dot{\phi}) = \frac{m}{2} \left(\frac{ds}{dt} \right)^2 = \frac{m}{2} \left\{ \left[1 + \left(\frac{dz}{dr} \right)^2 \right] \dot{r}^2 + r^2 \dot{\phi}^2 \right\}$$

(c) Lagrange's equations are

$$\frac{d}{dt} \left\{ \left[1 + \left(\frac{dz}{dr} \right)^2 \right] \dot{r} \right\} - \frac{1}{2} \frac{\partial}{\partial r} \left\{ \left[1 + \left(\frac{dz}{dr} \right)^2 \right] \dot{r}^2 \right\} - r\dot{\phi}^2 = 0 \qquad ; \; r\text{-eqn}$$

$$\frac{d}{dt} \left(r^2 \dot{\phi} \right) = 0 \qquad ; \; \phi\text{-eqn}$$

(d) If one takes the generalized coordinates to be $q^i = (r, \phi)$, with z determined by $z = z(r)$, then the metric and its inverse are

$$\begin{pmatrix} g_{rr} & g_{r\phi} \\ g_{\phi r} & g_{\phi\phi} \end{pmatrix} = \begin{pmatrix} 1 + (dz/dr)^2 & 0 \\ 0 & r^2 \end{pmatrix}$$

$$\begin{pmatrix} g^{rr} & g^{r\phi} \\ g^{\phi r} & g^{\phi\phi} \end{pmatrix} = \begin{pmatrix} 1/[1 + (dz/dr)^2] & 0 \\ 0 & 1/r^2 \end{pmatrix}$$

---

[5] *Note*: This problem is instructive as to the role of the metric in relating the coordinate motion to the actual physical motion.

(e) We sketch the situation in Fig. 5.1.

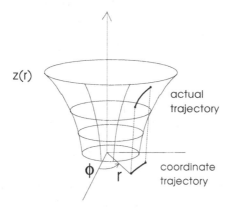

z(r)

actual
trajectory

coordinate
trajectory

φ   r

Fig. 5.1   Sketch of coordinates $q^i = (r, \phi)$ for particle moving without friction on surface of revolution $z(r)$. We compare a coordinate trajectory with the actual trajectory on the surface.

Problem 5.11 In Prob. 5.10:

(a) What is the affine connection on the surface?

(b) What is the Riemann tensor on the surface?

Solution to Problem 5.11

(a) There is no ϕ-dependence in the diagonal metric in Prob. 5.10(d), and all derivatives w.r.t. ϕ vanish. The only non-vanishing parts of the affine connection on the surface are then

$$\Gamma^{\phi}_{\phi r} = \Gamma^{\phi}_{r\phi} = \frac{1}{2}g^{\phi\phi}\left[\frac{\partial g_{\phi\phi}}{\partial r} + \frac{\partial g_{\phi r}}{\partial \phi} - \frac{\partial g_{r\phi}}{\partial \phi}\right] = \frac{1}{r}$$

$$\Gamma^{r}_{\phi\phi} = \frac{1}{2}g^{rr}\left[\frac{\partial g_{r\phi}}{\partial \phi} + \frac{\partial g_{r\phi}}{\partial \phi} - \frac{\partial g_{\phi\phi}}{\partial r}\right] = -\frac{1}{1 + (dz/dr)^2}\,r$$

$$\Gamma^{r}_{rr} = \frac{1}{2}g^{rr}\left[\frac{\partial g_{rr}}{\partial r} + \frac{\partial g_{rr}}{\partial r} - \frac{\partial g_{rr}}{\partial r}\right] = \frac{1}{1 + (dz/dr)^2}\frac{dz}{dr}\frac{d^2z}{dr^2}$$

$$; \text{all others vanish}$$

As a check, for constant z this must reduce to the plane polar coordinate results in Eqs. (3.84)

$$z = \text{constant} \qquad ; \text{plane polar coordinates}$$

(b) The symmetry properties imply that the only relevant component of the Riemann tensor is $R^r_{\ \phi r \phi}$. The affine connection has no dependence on ϕ. Hence, the Riemann tensor on the surface of revolution is given by

$$R^r_{\ \phi r \phi} = \left(\frac{\partial}{\partial r} \Gamma^r_{\phi\phi} + \Gamma^r_{rm} \Gamma^m_{\phi\phi} \right) - \left(\frac{\partial}{\partial \phi} \Gamma^r_{\phi r} + \Gamma^r_{\phi m} \Gamma^m_{\phi r} \right)$$

$$= -\frac{1}{1 + (dz/dr)^2} + \frac{2r}{[1 + (dz/dr)^2]^2} \left(\frac{dz}{dr} \right) \left(\frac{d^2 z}{dr^2} \right)$$

$$- \frac{r}{[1 + (dz/dr)^2]^2} \left(\frac{dz}{dr} \right) \left(\frac{d^2 z}{dr^2} \right) + \frac{1}{1 + (dz/dr)^2}$$

$$= \frac{r}{[1 + (dz/dr)^2]^2} \left(\frac{dz}{dr} \right) \left(\frac{d^2 z}{dr^2} \right)$$

Again, as a check, Probs. 5.5–5.6 imply that the Riemann tensor vanishes for constant z and plane polar coordinates.

Problem 5.12 The change in coordinate volume under the point transformation in Eqs. (3.26) is given by

$$d\xi^1 \cdots d\xi^n = \frac{\partial(\xi^1, \cdots, \xi^n)}{\partial(q^1, \cdots, q^n)} dq^1 \cdots dq^n$$

where the jacobian determinant is defined by

$$\frac{\partial(\xi^1, \cdots, \xi^n)}{\partial(q^1, \cdots, q^n)} \equiv \begin{vmatrix} \partial\xi^1/\partial q^1 & \cdots & \partial\xi^1/\partial q^n \\ \vdots & & \vdots \\ \partial\xi^n/\partial q^1 & \cdots & \partial\xi^n/\partial q^n \end{vmatrix}$$

(a) Define a matrix $\left(\underline{A}^{-1} \right)_{ij} \equiv \bar{a}^i_{\ j}$. Show

$$d\xi^1 \cdots d\xi^n = \left(\det \underline{A}^{-1} \right) dq^1 \cdots dq^n$$

(b) Show the corresponding change in metric can be written as a matrix relation

$$\bar{g} = (\underline{A})^T \underline{g} \, \underline{A}$$

Hence conclude the new determinant of the metric is

$$\det \bar{g} = (\det \underline{A})^2 \det g$$

(c) Show

$$(\det \underline{A}) (\det \underline{A}^{-1}) = 1$$

and hence conclude that the physical n-dimensional volume element in Eq. (5.152) is a *scalar* under coordinate transformations.[6]

Solution to Problem 5.12

(a) Assume an analytic *point transformation* between the new set of generalized coordinates $(\xi^1, \cdots, \xi^n)$ and the original set $(q^1, \cdots, q^n)$ of the form in Eqs. (3.26)

$$\xi^1 = \xi^1(q^1, \cdots, q^n)$$

$$\vdots$$

$$\xi^n = \xi^n(q^1, \cdots, q^n)$$

The corresponding coordinate transformation matrix is given in Eq. (3.29)

$$\bar{a}^i{}_j \equiv \frac{\partial \xi^i}{\partial q^j}$$

It is a result from analysis that the coordinate volumes are related by

$$d\xi^1 \cdots d\xi^n = \frac{\partial(\xi^1, \cdots, \xi^n)}{\partial(q^1, \cdots, q^n)} \, dq^1 \cdots dq^n$$

where the jacobian determinant is given by

$$\frac{\partial(\xi^1, \cdots, \xi^n)}{\partial(q^1, \cdots, q^n)} \equiv \begin{vmatrix} \partial \xi^1/\partial q^1 & \cdots & \partial \xi^1/\partial q^n \\ \vdots & & \vdots \\ \partial \xi^n/\partial q^1 & \cdots & \partial \xi^n/\partial q^n \end{vmatrix}$$

Define the matrix $\underline{A}^{-1}$ by

$$\left(\underline{A}^{-1}\right)_{ij} \equiv \bar{a}^i{}_j$$

Then by inspection

$$d\xi^1 \cdots d\xi^n = \left(\det \underline{A}^{-1}\right) dq^1 \cdots dq^n$$

The corresponding inverse matrix $\underline{A}$ follows from the orthogonality statement in Eqs. (3.32)

$$\bar{a}^i{}_j a^j{}_k = \delta^i{}_k$$

$$a^i{}_j \bar{a}^j{}_k = \delta^i{}_k$$

---

[6]The more astute reader can probably convert this into a proof that, within an overall constant, the n-dimensional volume element *must* have the form in Eq. (5.152).

It is given by

$$(\underline{A})_{jk} \equiv a^j{}_k = \bar{a}_k{}^j$$

where the final relation follows from Eqs. (3.36). Evidently

$$\underline{A}\,\underline{A}^{-1} = \underline{1}$$

(b) The metric transforms according to

$$\bar{g}_{ij} = \bar{a}_i{}^k \bar{a}_j{}^l\, g_{kl} = (\underline{A})_{ki}\, g_{kl}\, (\underline{A})_{lj}$$

In matrix language, this is

$$\bar{g} = (\underline{A})^T g \underline{A}$$

We make use of the following properties of the determinant:

- The determinant of a matrix product is the corresponding product of determinants;
- The determinant of the transpose of a matrix is the determinant of the matrix.

It follows that

$$\det \bar{g} = (\det \underline{A})^2 \det g$$

It also follows from the above that

$$(\det \underline{A})\left(\det \underline{A}^{-1}\right) = 1$$

(c) The physical n-dimensional volume element in Eq. (5.152) can therefore be written

$$
\begin{aligned}
d^{(n)}\tau &= \sqrt{g}\, dq^1 \cdots dq^n \qquad\qquad g = \det g\\
&= \sqrt{\bar{g}}\, (\det \underline{A})^{-1}\, dq^1 \cdots dq^n\\
&= \sqrt{\bar{g}}\left(\det \underline{A}^{-1}\right) dq^1 \cdots dq^n\\
&= \sqrt{\bar{g}}\, d\xi^1 \cdots d\xi^n
\end{aligned}
$$

The volume element $d^{(n)}\tau$ therefore transforms as a *scalar* under the coordinate transformation.

Chapter 6

Special Relativity

Problem 6.1 Given cartesian coordinates $(x^1, \cdots, x^4) = (x^1, x^2, x^3, ct)$ and the Lorentz metric of Eq. (6.5), show the affine connection and Riemann curvature tensor both vanish. One says that such a Minkowski space is "flat."

Solution to Problem 6.1

The Lorentz metric in Eq. (6.5) is

$$\underline{g} = \begin{bmatrix} 1 & 0 & 0 & 0 \\ 0 & 1 & 0 & 0 \\ 0 & 0 & 1 & 0 \\ 0 & 0 & 0 & -1 \end{bmatrix} \qquad ; \text{ Lorentz metric in SR}$$

From Prob. 5.4, the affine connection is

$$\Gamma^\lambda_{\mu\nu} = \frac{1}{2}g^{\lambda\sigma}\left[\frac{\partial g_{\sigma\mu}}{\partial q^\nu} + \frac{\partial g_{\sigma\nu}}{\partial q^\mu} - \frac{\partial g_{\mu\nu}}{\partial q^\sigma}\right] \qquad ; (\lambda, \mu, \nu, \sigma) = 1, 2, 3, 4$$

Since all the elements of the Lorentz metric are global constants, the derivatives all vanish, and hence

$$\Gamma^\lambda_{\mu\nu} = 0$$

From Prob. 5.4, the Riemann curvature tensor is

$$R^\lambda_{\ \mu\sigma\nu} = \left(\frac{\partial}{\partial q^\sigma}\Gamma^\lambda_{\mu\nu} + \Gamma^\lambda_{\sigma\rho}\Gamma^\rho_{\mu\nu}\right) - (\sigma \leftrightarrow \nu) \qquad ; (\lambda, \mu, \nu, \sigma) = 1, 2, 3, 4$$

This clearly also vanishes

$$R^\lambda_{\ \mu\sigma\nu} = 0$$

This Minkowski space is "flat."

Problem 6.2 Suppose one were to augment the lagrangian L in Eq. (6.36) with a static potential about the origin $V(\vec{x})$ so that

$$L = -mc^2 \left[1 - \frac{1}{c^2} \left(\frac{d\vec{x}}{dt} \right)^2 \right]^{1/2} - V(\vec{x})$$

Use Lagrange's equations and the hamiltonian to determine how Eqs. (6.30) are modified. Discuss.

Solution to Problem 6.2

With the cartesian coordinates $q^i = (x_1, x_2, x_3)$, Lagrange's equations read[1]

$$\frac{d}{dt} \frac{\partial L}{\partial (\partial x_i / \partial t)} = \frac{\partial L}{\partial x_i} \qquad ; i = 1, 2, 3$$

The augmented lagrangian is

$$L = -mc^2 \left[1 - \frac{1}{c^2} \left(\frac{d\vec{x}}{dt} \right)^2 \right]^{1/2} - V(\vec{x})$$

The canonical momentum is

$$p_i = \frac{\partial L}{\partial (\partial x_i / \partial t)} = \frac{m}{(1 - \beta^2)^{1/2}} \frac{dx_i}{dt} \qquad ; \vec{\beta} = \frac{\vec{v}}{c}$$

Lagrange's equations then read

$$\frac{d\vec{p}}{dt} = \frac{d}{dt} \left[\frac{m\vec{v}}{(1 - \beta^2)^{1/2}} \right] = -\vec{\nabla} V$$

The hamiltonian is given by

$$H = p_i \frac{dx_i}{dt} - L = \frac{mc^2 \beta^2}{(1 - \beta^2)^{1/2}} + mc^2 (1 - \beta^2)^{1/2} + V(\vec{x})$$

$$= \frac{mc^2}{(1 - \beta^2)^{1/2}} + V$$

If the hamiltonian has no explicit dependence on the time, it is a constant of the motion, in this case the total energy

$$\frac{dE}{dt} = \frac{d}{dt} \left[\frac{mc^2}{(1 - \beta^2)^{1/2}} + V \right] = 0$$

---

[1] For the classical mechanics, see [Fetter and Walecka (2003)].

These provide the appropriate generalization of Eqs. (6.30) in the presence of an additional static potential $V(\vec{x})$. The relativistic extension of Newton's second law now contains the force $-\vec{\nabla}V$, and the relativistic energy is now augmented by the potential.

Problem 6.3 Consider a particle of energy $E_L^2 = (mc^2)^2 + (c\vec{p}_L)^2$ incident on a target of mass M in the laboratory frame. Let $\mathbf{p} = (\vec{p}_L, E_L/c)$ be the four-momentum of the incident particle in that frame and $\mathbf{P} = (\vec{0}, Mc)$ that of the target. Express the Lorentz invariant $s \equiv -(\mathbf{p} + \mathbf{P})^2$ in both the laboratory and center-of-momentum (C-M) frames, and hence determine the total energy available in the C-M system in terms of E_L. The C-M frame is that one where the four-momenta are $\mathbf{p} = (\vec{p}, E_1/c)$ and $\mathbf{P} = (-\vec{p}, E_2/c)$, respectively.

Solution to Problem 6.3

Since the total momentum $\vec{p}+\vec{P}$ vanishes in the C-M system, the Lorentz invariant $s \equiv -(\mathbf{p} + \mathbf{P})^2$ expressed in that frame becomes

$$s = \frac{1}{c^2}(E_1 + E_2)^2 \equiv \frac{1}{c^2}W^2 \qquad ; \text{ C-M frame}$$

where W is the total energy in the C-M frame.

In the laboratory frame, with an incident energy $E_L^2 = p_L^2 c^2 + m^2 c^4$ on a target of mass M at rest, the invariant s becomes

$$s = -\vec{p}_L^2 + \frac{1}{c^2}(E_L + Mc^2)^2$$
$$= m^2 c^2 + M^2 c^2 + 2ME_L \qquad ; \text{ Lab frame}$$

If these two expressions are equated, one obtains

$$W^2 = 2(Mc^2)E_L + (Mc^2)^2 + (mc^2)^2$$

This expresses the total energy in the C-M frame in terms of the incident energy of a particle of rest-mass m on a stationary target of rest mass M.

Problem 6.4 (a) Construct a logical argument to show that if an observer in f sees the frame f' moving with velocity $\vec{v}$, then an observer in f' will see f moving with $-\vec{v}$.

(b) Use the Lorentz transformation of Eqs. (6.93), corresponding to the configuration in Fig. 6.10 in the text, to compute the motion of the origin O as viewed from the moving frame f'. Show $\bar{z} = -v\bar{t}$ and hence conclude that the frame f indeed appears to move with a velocity $-\vec{v}$ when viewed from the frame f'.

(c) Construct a logical argument to show that transverse spatial vectors should be unchanged under a Lorentz transformation.

Solution to Problem 6.4

(a) Consider two frames (f, f') where, as observed from f, the frame f' is moving away with velocity $\vec{v}$. Now simply rotate the whole configuration by $180°$ about an axis perpendicular to $\vec{v}$ and half-way between the frames. The role of the two frames is then interchanged to (f', f), and it is now exactly the same physical situation, only *reoriented* in space. For an observer at rest in f', the frame f appears to be moving away with $-\vec{v}$.

(b) The Lorentz transformation to go from coordinates (z, ct) in f to $(\bar{z}, c\bar{t})$ in f' (see Fig. 6.10 in the text) is given in Eqs. (6.93)

$$\bar{z} = \frac{z - vt}{(1 - \beta^2)^{1/2}}$$

$$\bar{t} = \frac{t - vz/c^2}{(1 - \beta^2)^{1/2}}$$

The origin O of f has coordinates $(0, ct)$. The corresponding coordinates $(\bar{z}, c\bar{t})$ of O as seen in f' are then

$$\bar{z} = \frac{-vt}{(1 - \beta^2)^{1/2}}$$

$$\bar{t} = \frac{t}{(1 - \beta^2)^{1/2}}$$

Hence

$$\bar{z} = -v\bar{t}$$

and the origin O appears to move with velocity $-v$ in the frame f'.

(c) Consider a rod and an opening of the same proper length transverse to a closing relative velocity $\vec{v}$. Suppose the transverse length of an object, when viewed from a rest frame, were to *shrink* at a finite velocity.[2] Then, as viewed from the rest frame of the opening, the rod would pass right through; however, when viewed from the rest frame of the rod, it would hit the opening. Now the rod either does, or does not, pass through the opening. This presents a contradiction, and hence the transverse length must be unaffected by the Lorentz transformation [see Eqs. (6.93)].

---

[2]One arrives at the same contradiction if it were to *grow*.

Problem 6.5 It was Lorentz who first showed that the coordinate transformation from (z, t) to $(\bar{z}, \bar{t})$ in Eq. (6.93) leaves the electromagnetic wave operator invariant (same c)

$$\frac{\partial^2}{\partial z^2} - \frac{1}{c^2}\frac{\partial^2}{\partial t^2} = \frac{\partial^2}{\partial \bar{z}^2} - \frac{1}{c^2}\frac{\partial^2}{\partial \bar{t}^2}$$

Verify this.

It was Einstein who gave this result a physical interpretation as the actual transformation of these quantities between inertial frames, which revolutionized our understanding of space-time.

Solution to Problem 6.5[3]

Lorentz observed that there is a *mathematical transformation* that leaves the form of the wave equation for light unchanged (leaves it "invariant"). That transformation is

$$z = \frac{(z' + Vt')}{\sqrt{1 - V^2/c^2}} \qquad ; \text{Lorentz transformation}$$

$$t = \frac{(t' + Vz'/c^2)}{\sqrt{1 - V^2/c^2}}$$

where V is simply some constant. These equations are readily inverted, and the result is obtained by merely changing the sign of $V \leftrightarrow -V$

$$z' = \frac{(z - Vt)}{\sqrt{1 - V^2/c^2}} \qquad ; \text{Lorentz transformation}$$

$$t' = \frac{(t - Vz/c^2)}{\sqrt{1 - V^2/c^2}}$$

These are Eqs. (6.93). It is readily verified that this transformation leaves the following quadratic form invariant

$$z^2 - c^2 t^2 = z'^2 - c^2 t'^2 \qquad ; \text{invariant}$$

Let us verify Lorentz's result. Write $\phi[z(z', t'), t(z', t')]$, and use the chain rule for differentiation of an implicit function twice. This gives

$$\frac{\partial \phi}{\partial z'} = \frac{\partial \phi}{\partial z}\frac{\partial z}{\partial z'} + \frac{\partial \phi}{\partial t}\frac{\partial t}{\partial z'} = \frac{1}{\sqrt{1 - V^2/c^2}}\left[\frac{\partial \phi}{\partial z} + \frac{V}{c^2}\frac{\partial \phi}{\partial t}\right]$$

$$\frac{\partial^2 \phi}{\partial z'^2} = \frac{1}{1 - V^2/c^2}\left\{\frac{\partial}{\partial z}\left[\frac{\partial \phi}{\partial z} + \frac{V}{c^2}\frac{\partial \phi}{\partial t}\right] + \frac{V}{c^2}\frac{\partial}{\partial t}\left[\frac{\partial \phi}{\partial z} + \frac{V}{c^2}\frac{\partial \phi}{\partial t}\right]\right\}$$

---

[3]The solution to Prob. 6.5 is taken from [Walecka (2008)]. Here V is a velocity.

In a similar fashion, one obtains

$$\frac{\partial \phi}{\partial t'} = \frac{\partial \phi}{\partial z}\frac{\partial z}{\partial t'} + \frac{\partial \phi}{\partial t}\frac{\partial t}{\partial t'} = \frac{1}{\sqrt{1 - V^2/c^2}}\left[V\frac{\partial \phi}{\partial z} + \frac{\partial \phi}{\partial t}\right]$$

$$\frac{\partial^2 \phi}{\partial t'^2} = \frac{1}{1 - V^2/c^2}\left\{V\frac{\partial}{\partial z}\left[V\frac{\partial \phi}{\partial z} + \frac{\partial \phi}{\partial t}\right] + \frac{\partial}{\partial t}\left[V\frac{\partial \phi}{\partial z} + \frac{\partial \phi}{\partial t}\right]\right\}$$

Now take the previous relation and subtract $(1/c^2)$ times this relation to arrive at

$$\frac{\partial^2 \phi}{\partial z'^2} - \frac{1}{c^2}\frac{\partial^2 \phi}{\partial t'^2} = \frac{\partial^2 \phi}{\partial z^2} - \frac{1}{c^2}\frac{\partial^2 \phi}{\partial t^2} \qquad ; \text{ Lorentz}$$

This is the result of Lorentz.

Problem 6.6 Consider all inertial frames whose origins coincide, but which are moving at different velocities $\vec{v}$. An event occurs in one frame at a space-time point $(\vec{x}, ct)$.

(a) Show that in all such frames, the event lies on a hyperboloid defined by $\vec{x}^2 - (ct)^2 = \text{constant}$ [see Fig. (6.1)].

(b) Show that all events connected by a light signal to the origin will give $\vec{x}^2 - (ct)^2 = 0$; they lie on the *light cone*. Hence conclude that all observers measure the same speed of light.

(c) Show that all events which stand in a causal relationship in one frame will preserve this relationship in all frames.

Solution to Problem 6.6

(a) Consider two frames moving with constant relative velocity $\vec{v}$ whose origins coincide $(\bar{x}, \bar{t}) = (x, t) = 0$. The coordinates in the two frames of a subsequent event are related by the Lorentz transformation in Eq. (6.91)

$$\bar{x}^\mu = \bar{a}^\mu{}_{\mu'} x^{\mu'}$$

The interval w.r.t. the origin is

$$\bar{x}_\mu \bar{x}^\mu = \bar{a}^\mu{}_{\mu'} \bar{a}_\mu{}^{\nu'} x^{\mu'} x_{\nu'}$$

Since

$$\bar{a}^\mu{}_{\mu'} \bar{a}_\mu{}^{\nu'} = a_{\mu'}{}^\mu \bar{a}_\mu{}^{\nu'} = \delta_{\mu'}{}^{\nu'}$$

the interval is preserved under the Lorentz transformation

$$\bar{x}_\mu \bar{x}^\mu = x_{\mu'} x^{\mu'}$$

This defines a hyberboloid

$$\vec{x}^2 - c^2 t^2 = \text{constant} \qquad ; \text{event hyperboloid}$$

We illustrate this hyperboloid in the case of two space and one time dimensions, with $x^\mu = (x^2, x^3, ct)$, in Fig. 6.1, where a causal relation between the two events, with $ct \geq |\vec{x}|$, is assumed so that the interval is time-like

$$\vec{x}^2 - c^2 t^2 < 0 \qquad ; \text{time-like}$$

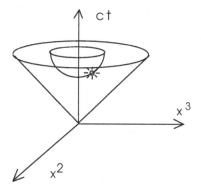

Fig. 6.1   Forward light-cone and event hyperboloid with a time-like interval.

No matter in which Lorentz frame the second event is viewed, it will lie somewhere on this "event hyperboloid".

(b) If the second event can be connected to the origin by a light signal. then $ct = |\vec{x}|$, and

$$\vec{x}^2 - c^2 t^2 = 0 \qquad ; \text{light cone}$$

This defines the "light cone", and the forward light cone with $t \geq 0$ is also shown in Fig. 6.1. Since this interval is again preserved under a Lorentz transformation, all observers will measure the *same speed of light*.

(c) Provided the sign of the time is not reversed under the proper, homogeneous Lorentz transformation, an event on the time-like event hyperboloid will always lie within the forward light cone in any Lorentz frame, and hence all events which stand in a causal relationship with $t > 0$ in one frame will preserve this relationship in all frames.[4]

Problem 6.7 A light signal, and a neutrino with energy $E = 2\,\text{MeV}$ and rest mass $mc^2 = 2\,\text{eV}$, are emitted simultaneously from a supernova which is at a distance 10^4 light-years from earth.
  (a) What is the difference in arrival times at the earth?
  (b) How long does the trip take in the neutrino's rest frame?
  (c) What is the distance to earth as viewed in the neutrino's rest frame?

Solution to Problem 6.7[5]

(a) As in the next problem, we use the ratio between the rest energy of the neutrino and its total energy to obtain its velocity

$$\frac{E_{\text{rest}}}{E} = \sqrt{1 - \frac{v^2}{c^2}} = \frac{2\,\text{eV}}{2 \times 10^6\,\text{eV}} = 10^{-6}$$

and therefore

$$v = c\sqrt{1 - 10^{-12}}$$

Call l the distance of the supernova from earth. The difference in the arrival times at earth is then given by[6]

$$\Delta t = \frac{l}{v} - \frac{l}{c} = \frac{l}{c}\left(\frac{1}{\sqrt{1 - 10^{-12}}} - 1\right) \approx 10^{-12}\frac{l}{2c}$$

One light-year is the distance traveled by light in a year

$$1\,\text{yr} = 3.15 \times 10^7\,\text{sec}$$
$$1\,\text{light-yr} = 3.15c \times 10^7\,\text{sec}$$

Hence

$$l = 10^4\,\text{light-yr} = 3.15c \times 10^{11}\,\text{sec}$$
$$\Delta t = 0.157\,\text{sec}$$

(b) Equation (6.60) provides the relation between the time intervals measured in the laboratory frame (t) and in the rest frame of the particle

---

[4]Note that this is *not* the case if the interval is space-like

$$\vec{x}^2 - c^2 t^2 > 0 \qquad\quad ; \text{ space-like}$$

You are urged to sketch this situation for yourself.
[5]The solutions to Probs. 6.7–6.8 are taken from [Amore and Walecka (2013)].
[6]Use $(1 + x)^{-1/2} = 1 - x/2 + 3x^2/8 + \cdots$.

(τ). There is a time dilation of the first with respect to the second and therefore

$$\tau = \frac{l}{v}\sqrt{1 - \frac{v^2}{c^2}} \approx \frac{l}{c}\sqrt{1 - \frac{v^2}{c^2}} = 10^{-2}\,\text{yr}$$

which is just 3.65 days.

(c) From part (a) of the problem, we know that the distance to earth viewed in the neutrino's rest frame will be 10^{-6} smaller than its actual value

$$l_{\text{rest}} = l\sqrt{1 - \frac{v^2}{c^2}} = 10^{-2}\,\text{light-yr}$$

Problem 6.8 An electron of energy 50 GeV and rest mass $mc^2 = 0.5$ MeV travels down an accelerator pipe of length 1 km. How long does the pipe appear to be in the rest frame of the electron (in cm)?

Solution to Problem 6.8

From the ratio between the rest energy and the total energy of this electron we can obtain $\sqrt{1 - v^2/c^2}$ [see Eqs. (6.31)]

$$\frac{E_{\text{rest}}}{E} = \sqrt{1 - \frac{v^2}{c^2}} = \frac{0.5\,\text{MeV}}{50 \times 10^3\,\text{MeV}} = 10^{-5}$$

We therefore conclude that the accelerator pipe will appear to the electron in its rest frame to be just 1 cm long

$$l_{\text{rest}} = \sqrt{1 - \frac{v^2}{c^2}}\,l = 10^{-5} \times 10^5\,\text{cm} = 1\,\text{cm}$$

Problem 6.9 Consider the local laboratory current of Eq. (6.113).

(a) Show that the number of particles in the spatial volume d^3x in the laboratory frame is given by $n(x)u^4(x)\,d^3x/c$. Then show that the particle flux through the surface $d\vec{S}$ is $n(x)\vec{u}(x) \cdot d\vec{S}$.

(b) The number of particles is conserved. Show this implies

$$\frac{1}{c}\frac{d}{dt}\int_V n(x)u^4(x)\,d^3x = -\oint_S n(x)\vec{u}(x) \cdot d\vec{S}$$

where V is a spatial laboratory volume and S is the surface surrounding V.

(c) Use Gauss's theorem and the fact that the relation in (b) must hold for any V to re-derive the laboratory continuity equation

$$\frac{\partial}{\partial x^\mu}\left[\frac{1}{c}n(x)u^\mu(x)\right] = 0$$

Solution to Problem 6.9

(a) Consider a small spatial volume $d^3\bar{x}$ in the local rest frame of the fluid. The number of particles in this volume is

$$dn = n\, d^3\bar{x} \qquad\qquad ; \text{ number of particles}$$

Imagine the volume is a small pillbox with faces transverse to the direction of motion. In the lab, this volume will then be Lorentz contracted to

$$d^3x = (1-\beta^2)^{1/2}d^3\bar{x}$$

Hence

$$dn = \frac{n}{(1-\beta^2)^{1/2}}\, d^3x$$

The four-velocity of the fluid for the situation illustrated in Fig. 6.12 in the text is given in Eq. (6.104)

$$\frac{1}{c}u^\mu = \frac{1}{(1-\beta^2)^{1/2}}[\vec{\beta},1]$$

Therefore

$$dn = n\frac{1}{c}u^4\, d^3x = N^4\, d^3x$$

where N^μ is given in Eq. (6.113). This is the desired result.

The particle density in the lab frame is evidently

$$n_{\text{lab}} = \frac{n}{(1-\beta^2)^{1/2}} \qquad\qquad ; \text{ particle density}$$

The particle current is then

$$n_{\text{lab}}\vec{v} = \frac{n\,\vec{v}}{(1-\beta^2)^{1/2}} = n\vec{u} \qquad\qquad ; \text{ particle current}$$

(b) Since the number of particles is conserved, the rate of decrease of the number of particles in a small volume V is equal to the particle flux out through the surface S enclosing that region

$$-\frac{1}{c}\frac{d}{dt}\int_V n(x)u^4(x)\,d^3x = \oint_S n(x)\vec{u}\,(x)\cdot d\vec{S}$$

Take the time derivative under the integral on the left, and use Gauss's theorem on the right, to give

$$-\int_V \frac{1}{c}\frac{\partial}{\partial t}[n(x)u^4(x)]\,d^3x = \int_V \vec{\nabla}\cdot[n(x)\vec{u}\,(x)]\,d^3x$$

(c) Since the volume V is arbitrarily small, the integrands in part (b) can be equated to give the laboratory continuity Eq. (6.122)

$$\frac{\partial}{\partial x^\mu}\left[\frac{1}{c}n(x)u^\mu(x)\right] = 0$$

Here $n(x)$ is now the local, rest-frame particle density, and $u^\mu(x)$ is the local, laboratory fluid four-velocity.

Problem 6.10 This problem retains the $O(1/c^2)$ corrections in taking the NRL of the relativistic expression for energy conservation in the fluid in Eq. (6.134), and leads to the result quoted in Eq. (6.138).

(a) Justify the following relation between the proper energy density ρc^2 and the laboratory internal energy density defined as $\rho\epsilon$ [compare Prob. 6.9(a)]

$$\rho c^2 = \rho\epsilon\,(1-\beta^2)^{1/2} = \rho\epsilon\left(1-\frac{1}{2}\beta^2+\cdots\right)$$

(b) Multiply Eq. (6.134) by c, and then expand in $1/c^2$ retaining all terms of overall $O(c^0)$ to obtain

$$\frac{\partial}{\partial t}\left(\rho c^2 + \rho v^2\right) + \vec{\nabla}\cdot\left[\vec{v}\left(\rho c^2 + \rho v^2 + P\right)\right] = 0$$

or ;
$$\frac{\partial}{\partial t}\left(\rho\epsilon + \frac{1}{2}\rho v^2\right) + \vec{\nabla}\cdot\left[\vec{v}\left(\rho\epsilon + \frac{1}{2}\rho v^2 + P\right)\right] = 0$$

If Eq. (6.123) is expanded to the same order, multiplied by mc^2, and subtracted from this expression, the result is to replace $\rho\epsilon \to \rho\epsilon'$ where $\rho\epsilon' = (\rho\epsilon - mc^2 n_{\text{lab}})$. The term $mc^2 n_{\text{lab}}$ is the rest mass contribution to the internal energy density in the laboratory frame. This difference, although still formally of $O(c^2)$, now makes the term $\rho\epsilon'$ comparable to the other terms, and this is the way the result is generally employed in the NRL.

(c) Verify that this is the result given in [Fetter and Walecka (2003)] p. 300 (where the possibility of an additional source term $\rho \vec{f}_{app} \cdot \vec{v}$ doing work on the system is also included). Verify also that Eq. (6.137) is the result given on p. 298 of that reference.

Solution to Problem 6.10

(a) Let ρc^2 be the energy density in the rest frame of the fluid (the "proper" energy density). The argument in Prob. 6.9(a) implies that due to the Lorentz contraction of the volume, the intrinsic energy density in the laboratory frame $\rho \epsilon$ is[7]

$$\rho \epsilon \equiv \rho_{lab} c^2 = \frac{\rho c^2}{(1 - \beta^2)^{1/2}}$$

Hence

$$\rho c^2 = \rho \epsilon (1 - \beta^2)^{1/2} = \rho \epsilon \left(1 - \frac{1}{2}\beta^2 + \cdots\right)$$

(b) Equation (6.134), multiplied by c, reads

$$\frac{\partial}{\partial t}(-P) + \frac{\partial}{\partial x^j}\left[\left(\rho + \frac{P}{c^2}\right)\frac{v^j c^2}{1 - \beta^2}\right] + \frac{\partial}{\partial t}\left[\left(\rho + \frac{P}{c^2}\right)\frac{c^2}{1 - \beta^2}\right] = 0$$

Now expand in $\beta^2 = v^2/c^2$ to obtain

$$\frac{\partial}{\partial t}\left[\rho c^2 + (\rho c^2 + P)(\beta^2 + \cdots)\right] + \frac{\partial}{\partial x^j}\left[(\rho c^2 + P)v^j(1 + \beta^2 + \cdots)\right] = 0$$

With the assumption

$$(\rho c^2 + P)\beta^2 \approx \rho v^2$$

in the correction terms, this becomes

$$\frac{\partial}{\partial t}\left(\rho c^2 + \rho v^2\right) + \vec{\nabla} \cdot \left[\vec{v}\left(\rho c^2 + \rho v^2 + P\right)\right] = 0$$

An iteration of the result in part (a) gives

$$\rho c^2 = \rho \epsilon - \frac{1}{2}\rho c^2 \beta^2 + \cdots = \rho \epsilon - \frac{1}{2}\rho v^2 + \cdots$$

---

[7]Picture $\rho \epsilon$ as arising from a set of finite-mass particles with internal excitations, where $\rho \epsilon = n_{lab}(mc^2 + \epsilon_{int})$. An additional kinetic-energy contribution of $\rho_{lab} v^2/2$ for this element is contained in the result derived below.

Thus in the NRL, the energy conservation Eq. (6.134) reduces to

$$\frac{\partial}{\partial t}\left(\rho\epsilon + \frac{1}{2}\rho v^2\right) + \vec{\nabla} \cdot \left[\vec{v}\left(\rho\epsilon + \frac{1}{2}\rho v^2 + P\right)\right] = 0 \qquad ; \text{NRL}$$

In addition to the intrinsic energy density in the laboratory frame $\rho\epsilon$, one has contributions from the kinetic energy density $\rho v^2/2$ and the pressure P.

After multiplication by c, the particle current conservation Eq. (6.123) reads

$$\frac{\partial}{\partial t}\left[\frac{n}{(1 - \beta^2)^{1/2}}\right] + \vec{\nabla} \cdot \left[\frac{n\vec{v}}{(1 - \beta^2)^{1/2}}\right] = 0$$

Identify (see Prob. 6.9)

$$n_{\text{lab}} = \frac{n}{(1 - \beta^2)^{1/2}}$$

Now multiply by the particle rest energy mc^2 to arrive at

$$\frac{\partial}{\partial t}\left[mc^2\, n_{\text{lab}}\right] + \vec{\nabla} \cdot \left[\vec{v}\left(mc^2\, n_{\text{lab}}\right)\right] = 0 \qquad ; \text{current conservation}$$

If this expression is subtracted from the result in part (b), as explained in the statement of the problem, one obtains

$$\frac{\partial}{\partial t}\left(\rho\epsilon' + \frac{1}{2}\rho v^2\right) + \vec{\nabla} \cdot \left[\vec{v}\left(\rho\epsilon' + \frac{1}{2}\rho v^2 + P\right)\right] = 0$$

$$\rho\epsilon' \equiv \rho\epsilon - mc^2\, n_{\text{lab}}$$

where the new laboratory energy density $\rho\epsilon'$ now excludes the rest mass.

(c) This is the result given in the discussion of NRL hydrodynamics in [Fetter and Walecka (2003)], on p. 300, where the possibility of an additional source term $\rho\vec{f}_{\text{app}} \cdot \vec{v}$ doing work on the system is also included.

It is also readily verified that Eq. (6.137) is the basic equation of hydrodynamics in the NRL, given on p. 298 of that reference.

Problem 6.11 The energy-momentum tensor for a fluid without any shear forces is given by Eq. (6.108), where the four-velocity is obtained from Eq. (6.104). How is the stress tensor modified by the inclusion of shear forces? (*Hint:* see [Fetter and Walecka (2003)] chap. 12 for a discussion of viscosity, and [Weinberg (1972)] for the relativistic generalization.)

Solution to Problem 6.11

In the discussion of the viscosity of a non-relativistic fluid in chapter 12 of [Fetter and Walecka (2003)], it is stated that the stress (energy-momentum) tensor should be symmetric (to produce a conserved angular momentum tensor) and linear in the velocity gradients. Thus the following viscous contribution is added to the non-relativistic form of the tensor in Eq. (6.108)

$$T_{ij}^{\text{vis}} = -\eta \left(\frac{\partial v_i}{\partial x_j} + \frac{\partial v_j}{\partial x_i} - \frac{2}{3}\delta_{ij}\vec{\nabla} \cdot \vec{v} \right) - \zeta\delta_{ij}\left(\vec{\nabla} \cdot \vec{v}\right)$$

where η and ζ are two phenomenological constants known as the *viscosity* and the *bulk* (or *second*) *viscosity*. It is shown in [Fetter and Walecka (2003)] that this leads to the following non-relativistic hydrodynamic equation for a viscous fluid [compare Eq. (6.137)]

$$\frac{\partial \vec{v}}{\partial t} + (\vec{v} \cdot \vec{\nabla})\vec{v} = \vec{f}_{\text{app}} - \frac{1}{\rho}\vec{\nabla}P + \frac{\eta}{\rho}\nabla^2\vec{v} + \frac{1}{\rho}\left(\zeta + \frac{1}{3}\eta\right)\vec{\nabla}\left(\vec{\nabla} \cdot \vec{v}\right)$$

For an incompressible fluid with $\vec{\nabla} \cdot \vec{v} = 0$, this becomes the Navier-Stokes equation

$$\frac{\partial \vec{v}}{\partial t} + (\vec{v} \cdot \vec{\nabla})\vec{v} = \vec{f}_{\text{app}} - \frac{1}{\rho}\vec{\nabla}P + \nu\,\nabla^2\vec{v} \qquad ; \nu \equiv \frac{\eta}{\rho}$$

In special relativity, a shear tensor can be defined according to

$$W_{\mu\nu} \equiv \frac{\partial u_\mu}{\partial x^\nu} + \frac{\partial u_\nu}{\partial x^\mu} - \frac{2}{3}g_{\mu\nu}\left(\frac{\partial u^\lambda}{\partial x^\lambda}\right)$$

where u^μ is the four-velocity of the fluid [see Eq. (6.104) and Prob. 6.9]. An obvious relativistic generalization of the above is then

$$T_{\mu\nu}^{\text{vis}} = -\eta W_{\mu\nu} - \zeta g_{\mu\nu}\left(\frac{\partial u^\lambda}{\partial x^\lambda}\right)$$

In order to preserve the identification of ρ as the mass density in the rest frame of the fluid, as well as the statement of energy conservation there, [Weinberg (1972)] employs only a *projection* of this tensor. Define

$$h_{\mu\nu} \equiv g_{\mu\nu} + \frac{1}{c^2}u_\mu u_\nu$$

Then in the rest frame of the fluid, where the origin is at rest and we neglect terms of $O(v^2/c^2)$

$$h_{ij} = \delta_{ij} \qquad ; \ h_{4i} = h_{i4} = -v_i/c$$
$$h_{44} = 0$$

For the viscous contribution to the relativistic energy-momentum tensor of the fluid, which is to be added to Eq. (6.108), [Weinberg (1972)] uses

$$T^{\text{vis}}_{\mu\nu} = -\eta \left[h_{\mu\rho} W^{\rho\sigma} h_{\sigma\nu} \right] - \zeta \, h_{\mu\nu} \left(\frac{\partial u^\lambda}{\partial x^\lambda} \right)$$

In the rest frame of the fluid, neglecting $O(v^2/c^2)$ and $O[(v/c)\boldsymbol{\nabla}v]$, one then has[8]

$$T^{\text{vis}}_{44} = T^{\text{vis}}_{4i} = T^{\text{vis}}_{i4} = 0$$
$$T^{\text{vis}}_{ij} = -\eta \left(\frac{\partial v_i}{\partial x_j} + \frac{\partial v_j}{\partial x_i} - \frac{2}{3}\delta_{ij}\vec{\nabla}\cdot\vec{v} \right) - \zeta\delta_{ij}\left(\vec{\nabla}\cdot\vec{v} \right)$$

As justification for this result, we note the following:

- $T^{\text{vis}}_{\mu\nu}$ is certainly a second-rank tensor in special relativity;
- Since the additional contributions all vanish, the analysis of the covariant divergence $\partial T^{4\nu}/\partial x^\nu$ in Eqs. (6.130)–(6.134) remains unchanged. Hence, the statement of energy conservation in the NRL in Eq. (6.138) still holds [recall Prob. 6.10];
- For the analysis of $\partial T^{i\nu}/\partial x^\nu$, one now has an additional term $\partial T^{\text{vis}}_{ij}/\partial x_j$ on the l.h.s. of Eq. (6.133). This implies:
    - The relativistic Euler Eq. (6.136) now has an additional driving term;[9] however,
    - The viscous modification of the NRL in Eq. (6.137) is now exactly that detailed in [Fetter and Walecka (2003)], as described above!

- The particle current N_μ remains as analyzed in Prob. 6.9.

Problem 6.12 There is another representation of flat Minkowski space, which, while algebraically attractive, greatly obscures the transition from special to general relativity. In this approach, the basis vectors form an

---

[8]In the rest frame of the fluid the velocity vanishes at the origin, but the velocity *gradients* do not. For completeness, one should also include heat transfer and viscous heating {see [Fetter and Walecka (2003); Weinberg (1972)]}.

[9]We leave its derivation to the very dedicated reader.

ordinary four-dimensional orthonormal set with $\mathbf{e}_\mu \cdot \mathbf{e}_\nu = \delta_{\mu\nu}$, where $\delta_{\mu\nu}$ is the usual Kronecker delta. The metric is then simply the identity matrix, and there is no distinction between up and down indices. The fourth coordinate now becomes *imaginary* so that $x_\mu = (x_1, \cdots, x_4)$ with $x_4 = ix_0 = ict$. The square of this four-vector is then $\mathbf{x}^2 = x_1^2 + x_2^2 + x_3^2 + x_4^2 = \vec{x}^2 - c^2t^2$ and all of the effects of the negative fourth component in the metric are reproduced. In this approach, a Lorentz transformation is one that preserves the square of the four-vector.

(a) Show that a Lorentz transformation now forms an orthogonal matrix with $\underline{a}^T = \underline{a}^{-1}$.

(b) The Lorentz transformation corresponding to the configuration in Fig. 6.10 in the text is now a rotation

$$\bar{x}_3 = x_3 \cos\psi + x_4 \sin\psi$$
$$\bar{x}_4 = -x_3 \sin\psi + x_4 \cos\psi$$

From the motion of O' as observed in the frame f, show $\tan\psi = iv/c$.

(c) Re-derive the results for $(\bar{z}, \bar{t})$ in Eqs. (6.93).

(d) Exhibit the new Lorentz transformation matrix $\underline{\bar{a}}$ corresponding to that in Eqs. (6.93).

(e) Rewrite part (b) as a relation between real components

$$\bar{x}_3 = x_3 \cosh\chi + x_4' \sinh\chi$$
$$\bar{x}_4' = x_3 \sinh\chi + x_4' \cosh\chi$$

where $x_4' = -ix_4$ and $\tanh\chi = -\beta$.

Solution to Problem 6.12

(a) In this approach, a homogeneous Lorentz transformation leaves the length of the four-vector x_μ unchanged

$$x_\mu = a_{\mu\nu}x_\nu$$
$$x_\mu x_\mu = a_{\mu\rho}a_{\mu\sigma}x_\rho x_\sigma = x_\rho x_\rho$$

where repeated Greek indices are again summed from 1 to 4, and there is now no distinction between upper and lower indices. Hence

$$a_{\mu\rho}a_{\mu\sigma} = \delta_{\rho\sigma}$$

where $\delta_{\rho\sigma}$ is the usual Kronecker delta. In matrix language this is

$$\underline{a}^T \underline{a} = \underline{1}$$

Therefore

$$\underline{a}^T = \underline{a}^{-1}$$

A Lorentz transformation is a *rotation* in this complex Minkowski space.

(b) It follows that the Lorentz transformation corresponding to the configuration in Fig. 6.10 in the text is a rotation that mixes the (x_3, x_4) components

$$\bar{x}_3 = x_3 \cos\psi + x_4 \sin\psi$$
$$\bar{x}_4 = -x_3 \sin\psi + x_4 \cos\psi$$

The origin $(\bar{x}_3, \bar{x}_4) = (0, i c \bar{t})$ is the point $(x_3, x_4) = (vt, ict)$. Hence

$$0 = vt \cos\psi + ict \sin\psi$$
$$i c \bar{t} = -vt \sin\psi + ict \cos\psi$$

The first equation gives

$$\tan\psi = -\frac{v}{ic} = i\frac{v}{c}$$

Note this is a rotation through an imaginary angle. It follows that

$$\sin\psi = \frac{iv/c}{\sqrt{1 - v^2/c^2}} \qquad ; \ \cos\psi = \frac{1}{\sqrt{1 - v^2/c^2}}$$

(c) The full rotation in part (b) then becomes

$$\bar{z} = \frac{z - vt}{\sqrt{1 - v^2/c^2}}$$
$$\bar{t} = \frac{t - vz/c^2}{\sqrt{1 - v^2/c^2}}$$

These are Eqs. (6.93).

(d) The full Lorentz transformation matrix then becomes

$$\underline{a} = \frac{1}{(1 - \beta^2)^{1/2}} \begin{bmatrix} (1 - \beta^2)^{1/2} & 0 & 0 & 0 \\ 0 & (1 - \beta^2)^{1/2} & 0 & 0 \\ 0 & 0 & 1 & i\beta \\ 0 & 0 & -i\beta & 1 \end{bmatrix} \qquad ; \beta \equiv \frac{v}{c}$$

$$\underline{a}^T = \underline{a}^{-1}$$

(e) Use the following hyperbolic relations

$$\sin(i\chi) = i\sinh\chi \qquad ; \cos(i\chi) = \cosh\chi$$
$$\tan(i\chi) = i\tanh\chi$$

Define $\psi \equiv -i\chi$. The rotation in part (b) then becomes

$$\bar{x}_3 = x_3\cosh\chi - ix_4\sinh\chi \qquad ;\tanh\chi = -\frac{v}{c}$$
$$\bar{x}_4 = ix_3\sinh\chi + x_4\cosh\chi$$

Now write

$$x'_4 \equiv = -ix_4 = ct \qquad ; \bar{x}'_4 \equiv -i\bar{x}_4 = c\bar{t}$$

The above then becomes a relation between *real* components

$$\bar{x}_3 = x_3\cosh\chi + x'_4\sinh\chi \qquad ;\tanh\chi = -\frac{v}{c}$$
$$\bar{x}'_4 = x_3\sinh\chi + x'_4\cosh\chi$$

This reproduces Eqs. (6.93).

Problem 6.13 With the knowledge that the Lorentz transformation in Prob. 6.12(b) is a rotation, analyze the consequences of two successive Lorentz transformations. Show the velocities add according to

$$\beta_{\text{lab}} = \tanh\left[\tanh^{-1}(\beta_1) + \tanh^{-1}(\beta_2)\right]$$

and conclude that the laboratory velocity always satisfies $|\beta_{\text{lab}}| < 1$.

Solution to Problem 6.13

Two successive Lorentz transformations correspond to two successive rotations in Prob. 6.12, for which the rotation angles $\psi = i\chi$ add

$$\psi = \psi_1 + \psi_2$$
$$\text{or;} \qquad \chi = \chi_1 + \chi_2$$

Since $\tanh(-\chi) = -\tanh(\chi) = v/c$, this implies the following addition law for velocities

$$-\tanh^{-1}\left(\frac{v}{c}\right) = -\tanh^{-1}\left(\frac{v_1}{c}\right) - \tanh^{-1}\left(\frac{v_2}{c}\right)$$
$$\frac{v}{c} = \tanh\left[\tanh^{-1}\left(\frac{v_1}{c}\right) + \tanh^{-1}\left(\frac{v_2}{c}\right)\right]$$

The initial velocities satisfy the conditions

$$\left|\frac{v_1}{c}\right| < 1 \qquad ; \qquad \left|\frac{v_2}{c}\right| < 1$$

The function $\tanh(x)$ is plotted against x in Fig. 6.2; $\tanh(x)$ lies between -1 and 1, and it takes all values in between.

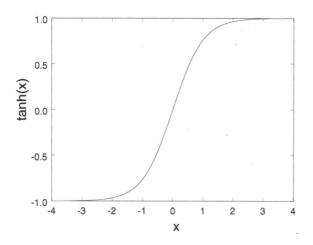

Fig. 6.2   The function $\tanh(x)$ vs. x.

The inverse function $\tanh^{-1}(u)$ follows immediately from this plot

$$\tanh^{-1}(u) = x$$
$$u = \tanh(x)$$

Find the appropriate $\tanh(x)$, and read off the value of x, which can take all values.

Then, whatever the value of the sum $[\tanh^{-1}(v_1/c) + \tanh^{-1}(v_2/c)]$ in the above, the hyperbolic tangent of this quantity will lie between -1 and 1. Hence the addition law for velocities with Lorentz transformations always results in a new velocity that satisfies the inequality

$$\left|\frac{v}{c}\right| < 1$$

Note that in the non-relativistic limit, where

$$\tanh(x) \approx \tanh^{-1}(x) \approx x \qquad ; \; x \to 0$$

we recover the newtonian result

$$v = v_1 + v_2 \qquad ; \; (v_1, v_2) \ll c$$

Problem 6.14 Consider the configuration in Fig. 6.10 in the text where a first event marks the coincidence of the origins. A light signal is emitted at that first event and received later at a second event at (z, t), with $z/t = c$. The coordinates of the second event in the second frame are $(\bar{z}, \bar{t})$, which are related to (z, t) by the Lorentz transformation in Eq. (6.93).

(a) Express $\bar{z}$ in terms of z and note the relation between them;
(b) Express $\bar{t}$ in terms of t and note the relation between them;
(c) Show the speed of light given by $\bar{z}/\bar{t}$ in the second frame is again c.

Solution to Problem 6.14

(a) The Lorentz transformation in Eqs. (6.93) is

$$\bar{z} = \frac{z - vt}{(1 - \beta^2)^{1/2}}$$

$$\bar{t} = \frac{t - vz/c^2}{(1 - \beta^2)^{1/2}}$$

The origins coincide for the first event, $(z, t) = (\bar{z}, \bar{t}) = 0$. The second event then has the spatial coordinates

$$z = ct \qquad ; \text{ second event}$$

$$\bar{z} = z \left(\frac{1 - \beta}{1 + \beta} \right)^{1/2}$$

where we have expressed $\bar{z}$ in terms of z. Note the reduction of the spatial distance.

(b) The times are related by

$$\bar{t} = t \left(\frac{1 - \beta}{1 + \beta} \right)^{1/2}$$

where we have similarly expressed $\bar{t}$ in terms of t. The time coordinate is also compressed.

(c) The ratio $\bar{z}/\bar{t} = z/t$ is preserved, verifying that the speed of light is identical in the two frames.

$$\frac{z}{t} = \frac{\bar{z}}{\bar{t}} = c$$

Problem 6.15 This problem concerns the Doppler shift in special relativity. A light signal is sent out from a source at $\bar{O}$ moving with velocity V in the z-direction. In its rest frame the source undergoes dN oscillations in a proper time $d\tau$. All observes can agree on this number dN. The proper frequency in the rest frame of the source is

$$\bar{\nu} = \frac{dN}{d\tau} \qquad ; \; dN = \bar{\nu}\, d\tau$$

Now Lorentz transform to the laboratory frame where the source is moving with velocity V.

(a) Show the corresponding time interval dt in the lab is given by the relation

$$d\tau = \left(1 - \frac{V^2}{c^2}\right)^{1/2} dt$$

(b) During the time dt, the light front has traveled a distance $dl = cdt$ in the laboratory frame and hence arrives at an origin O a distance $dl = cdt$ away as shown in Fig. 6.3(a).

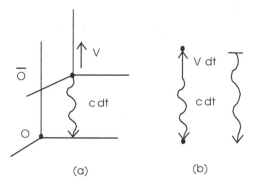

(a) (b)

Fig. 6.3 Configuration for calculation of the Doppler shift in special relativity.

In the time dt, the source has moved to a new position a distance $dl' = cdt + Vdt$ away from O. The wavelength of the radiation received at O is thus increased as indicated in Fig. 6.3(b). Since λ is the actual distance the signal has traveled divided by the number of oscillations, show that

$$\lambda = \frac{(c+V)dt}{dN} = \bar{\lambda}\,\frac{(1+V/c)}{(1-V^2/c^2)^{1/2}}$$

Here $\bar{\lambda} = c/\bar{\nu}$ is the proper wavelength of the radiation for a source at rest.

(c) Show that in the limit $V/c \to 0$, this reduces the familiar freshman physics result. Note that $V = \pm|V|$ can have either sign in these arguments.

Solution to Problem 6.15[10]

(a) Let us call O' and O the origins of two distinct frames: in the first frame the source is not moving, whereas in the second frame the source is moving with a constant velocity V in the z-direction. The two origins initially coincide.[11] After a given time the source has not moved in the first frame and therefore

$$z' = 0$$
$$t' = \tau$$

where τ is the proper time. In the second frame the coordinate of the source will be

$$z = Vt$$

where t is the time which has passed according to the observer's watch after the instant when the two origins coincided.

The coordinates of the two frames are related by a Lorentz transformation

$$z = \frac{(z' + Vt')}{\sqrt{1 - V^2/c}}$$
$$t = \frac{(t' + Vz'/c^2)}{\sqrt{1 - V^2/c}}$$

With the use of the above equations, we obtain the relation

$$t = \frac{\tau}{\sqrt{1 - V^2/c^2}}$$

In the present case, after making the identifications $\tau \to d\tau$ and $t \to dt$, we obtain the anticipated relation between the proper time in the rest frame, and the time interval in the laboratory frame where the source is moving with velocity V

$$d\tau = \sqrt{1 - V^2/c^2}\, dt$$

---

[10]The solution to Prob. 6.15 is taken from [Amore and Walecka (2013)].
[11]See Fig. 6.4 in the text.

(b) The configuration for the Doppler shift is illustrated in Fig. 6.3. λ is the wavelength of the signal received at O, and it is obtained as the actual distance the signal has traveled divided by the number of oscillations

$$\lambda = \frac{(c+V)dt}{dN} = \frac{(c+V)dt}{\bar{\nu}d\tau}$$

where $\bar{\nu}$ is the proper frequency of the signal, that is, the frequency in the frame where the source is at rest.

With the use of the proper wavelength of the signal $\bar{\lambda} = c/\bar{\nu}$, and the use of the result from part (a) that $d\tau = \sqrt{1 - V^2/c^2}\, dt$, we obtain the Doppler-shifted wavelength

$$\lambda = \bar{\lambda}\frac{(1+V/c)}{(1-V^2/c^2)^{1/2}} \qquad ; \text{ Doppler-shift}$$

(c) In the non-relativistic limit $V/c \to 0$ the above formula reduces to the freshman-physics result for the Doppler shift of a wave

$$\lambda \approx \bar{\lambda}\,(1+V/c) \qquad ; |V|/c \ll 1$$

The corresponding fractional shift in frequency will then be

$$\frac{\Delta\nu}{\bar{\nu}} \equiv \frac{1}{\bar{\nu}}\,(\nu - \bar{\nu}) = \frac{1}{\bar{\nu}}\left(\frac{c}{\lambda} - \frac{c}{\bar{\lambda}}\right) \approx -\frac{V}{c}$$

Note that in these arguments, V can have either sign.

Problem 6.16 A more elegant derivation of Eq. (6.105) is obtained as follows.[12]

(a) Show that Eqs. (6.102) and (6.103) imply that the combination $a^{\mu}{}_i a^{\nu}{}_i$ should transform as a symmetric second-rank Lorentz tensor. Then write this expression in terms of the only two such tensors available

$$a^{\mu}{}_i a^{\nu}{}_i = c_1 g^{\mu\nu} + c_2 u^{\mu} u^{\nu}$$

where (c_1, c_2) are Lorentz scalars.

(b) Use this relation to obtain two equations for (c_1, c_2), and hence rederive Eq. (6.105).

---

[12] Courtesy of Paolo Amore.

Solution to Problem 6.16

(a) Equation (6.102) expresses the energy-momentum tensor in terms of the Lorentz transformation in Eq. (6.82)

$$T^{\mu\nu} = P a^\mu{}_i a^\nu{}_i + \rho c^2 a^\mu{}_4 a^\nu{}_4$$

where the pressure and energy density $(P, \rho c^2)$ are Lorentz scalars, defined in the rest frame of the fluid. Equation (6.103) then expresses the last combination in terms of the four-velocity of the fluid

$$a^\mu{}_4 a^\nu{}_4 = \frac{1}{c^2} u^\mu u^\nu$$

$$\frac{1}{c} u^\mu = \frac{1}{(1-\beta^2)^{1/2}} [0, 0, \beta, 1]$$

Now both $T^{\mu\nu}$ and $u^\mu u^\nu / c^2$ are symmetric, second-rank tensors. It is then evident from the first relation above, that so is $a^\mu{}_i a^\nu{}_i$. This quantity can be written as a linear combination of the two symmetric tensors available

$$a^\mu{}_i a^\nu{}_i = c_1 g^{\mu\nu} + c_2 u^\mu u^\nu$$

where (c_1, c_2) are Lorentz scalars.

(b) One can obtain two relations for (c_1, c_2) by using particular values of (μ, ν), and a subset of Eqs. (6.106)

$$a^1{}_i a^1{}_i = 1 = c_1$$

$$a^3{}_i a^4{}_i = \frac{\beta}{1-\beta^2} = \frac{c^2 \beta}{1-\beta^2} c_2$$

It follows that

$$c_1 = 1 \qquad ; \; c_2 = \frac{1}{c^2}$$

Hence

$$a^\mu{}_i a^\nu{}_i = g^{\mu\nu} + \frac{1}{c^2} u^\mu u^\nu$$

This is Eq. (6.105).

Chapter 7

General Relativity

Problem 7.1 Given a four-dimensional riemannian space with an arbitrary metric. Take all the symmetry properties into account, and the first Bianchi identity, and determine the number of independent components of the Riemann curvature tensor R_{ijkl}.

Solution to Problem 7.1

The Riemann curvature tensor in four dimensions has $4^4 = 256$ components, It is antisymmetric in the first pair of indices (there are 6 possibilities), antisymmetric in the second pair, and symmetric in the interchange of the first and second pairs [see Eqs. (5.93)–(5.95) and (5.120)]. This reduces the number of independent components to 21, which we write out below

$$R_{1234} \quad ; R_{1232} \quad ; R_{1231} \quad ; R_{1242} \quad ; R_{1241} \quad ; R_{1343} \quad ; R_{1342}$$
$$R_{1341} \quad ; R_{1332} \quad ; R_{1443} \quad ; R_{1442} \quad ; R_{1432} \quad ; R_{2343} \quad ; R_{2342}$$
$$R_{3442} \quad ; R_{1212} \quad ; R_{1313} \quad ; R_{1414} \quad ; R_{2323} \quad ; R_{2424} \quad ; R_{3434}$$

The first Bianchi identity in Eq. (5.96) appears to provide one additional linear relation between these quantities, for any (lowered) k.[1] For example

$$R_{1234} + R_{1342} + R_{1423} = R_{1234} + R_{1342} - R_{1432} = 0$$

Other values of k yield the same relation. This reduces the number of independent components to 20.

---

[1] To get a non-trivial constraint, the indices (ilj) in Eq. (5.96) must be distinct, and different from k. For example, when the first Bianchi identity is applied to R_{1j1l}, one merely obtains $R_{1j1l} + R_{1l j1} = 0$, and when applied to R_{1jjl}, one obtains only $R_{1jjl} + R_{1jlj} = 0$.

Problem 7.2 Even though the Ricci tensor vanishes everywhere outside of a spherically symmetric source with the Schwarzschild solution, various individual components of the Riemann curvature tensor do not, and the space is curved. For example,

(a) Show that

$$R^4{}_{\theta 4\theta} = -\frac{rB'}{2AB} = -\frac{R_s}{2r}$$

(b) Find a small loop in a two-dimensional surface in the space where the enclosed surface manifests this curvature.

Solution to Problem 7.2

(a) From Eqs. (5.47) and Prob. 6.1, the Riemann curvature tensor in spherical coordinates is

$$R^\lambda{}_{\mu\sigma\nu} = \left(\frac{\partial}{\partial q^\sigma}\Gamma^\lambda{}_{\mu\nu} + \Gamma^\lambda{}_{\sigma\rho}\Gamma^\rho{}_{\mu\nu}\right) - (\sigma \leftrightarrow \nu) \qquad ; \ (\lambda, \sigma, \mu, \nu) = (r, \theta, \phi, ct)$$

$$= \left(\frac{\partial}{\partial q^\sigma}\Gamma^\lambda{}_{\mu\nu} + \Gamma^\lambda{}_{\sigma\rho}\Gamma^\rho{}_{\mu\nu}\right) - \left(\frac{\partial}{\partial q^\nu}\Gamma^\lambda{}_{\mu\sigma} + \Gamma^\lambda{}_{\nu\rho}\Gamma^\rho{}_{\mu\sigma}\right)$$

The affine connection with a static, spherically symmetric source and the metric of Eq. (7.47) is given in Table 7.1 in the text (and reproduced below in Prob. 7.3). The Ricci tensor $R_{\mu\nu} = R^\lambda{}_{\mu\lambda\nu}$, which vanishes outside the source, is calculated in the text. Here we compute the element

$$R^4{}_{\theta 4\theta} = \left(\frac{\partial}{\partial ct}\Gamma^4{}_{\theta\theta} + \Gamma^4{}_{4\rho}\Gamma^\rho{}_{\theta\theta}\right) - \left(\frac{\partial}{\partial \theta}\Gamma^4{}_{\theta 4} + \Gamma^4{}_{\theta\rho}\Gamma^\rho{}_{\theta 4}\right)$$

$$= \Gamma^4{}_{4r}\Gamma^r{}_{\theta\theta} = \left(\frac{B'}{2B}\right)\left(-\frac{r}{A}\right)$$

The Schwarzschild solution outside the source is given in Eqs. (7.88)

$$A(r) = \frac{1}{1 - (2MG/c^2 r)} \qquad ; \ B(r) = 1 - \frac{2MG}{c^2 r}$$

Hence we arrive at the following non-zero result for this particular element of the Riemann tensor outside a static spherically symmetric source

$$R^4{}_{\theta 4\theta} = -\frac{MG}{c^2 r} = -\frac{R_s}{2r}$$

This is the stated answer.

(b) Fix the coordinates (r, ϕ). One is then restricted to a 2-D surface in the 4-D space of coordinates (r, θ, ϕ, ct) about the source. Now consider *two events* separated by the following interval with the Schwarzschild metric

$$(ds)^2 = r^2(d\theta)^2 - \left(1 - \frac{2MG}{c^2 r}\right)(cdt)^2$$

Carry the pair around a closed loop which on the record keeper's screen appears as a rectangle with a horizontal length of cdt and vertical height of $rd\theta$. The area of this loop is $S^{4\theta} = (rd\theta)(cdt)$. Then

$$R^4{}_{\theta 4\theta} S^{4\theta} \neq 0$$

Thus, although the Ricci tensor vanishes, this expression is non-zero and the space is curved.[2]

Problem 7.3 Make use of the results in Table 7.1 in the text to write out the geodesic Eqs. (7.8) with coordinates $q^\mu = (r, \theta, \phi, ct)$ and a spherically symmetric source. Show that if one changes variable from proper time to time $(\tau \to t)$ the result is

$$\frac{1}{\rho}\frac{d}{dt}(\rho\dot{r}) + \left[\frac{A'}{2A}(\dot{r})^2 - \frac{r}{A}(\dot{\theta})^2 - \frac{r\sin^2\theta}{A}(\dot{\phi})^2 + \frac{B'}{2A}c^2\right] = 0$$

$$\frac{1}{\rho}\frac{d}{dt}(\rho\dot{\theta}) + \left[\frac{2}{r}(\dot{r}\dot{\theta}) - \sin\theta\cos\theta(\dot{\phi})^2\right] = 0$$

$$\frac{1}{\rho}\frac{d}{dt}(\rho\dot{\phi}) + \left[\frac{2}{r}(\dot{r}\dot{\phi}) + 2\frac{\cos\theta}{\sin\theta}(\dot{\theta}\dot{\phi})\right] = 0$$

$$\frac{1}{\rho}\frac{d}{dt}(\rho c) + \left[\frac{B'}{B}(c\dot{r})\right] = 0$$

Here the dot indicates a time derivative $(\dot{r} = dr/dt$, etc.), and ρ is given by

$$\rho = \left\{B - \frac{1}{c^2}\left[(\dot{r})^2 A + (r\dot{\theta})^2 + (r\dot{\phi}\sin\theta)^2\right]\right\}^{-1/2} = \frac{dt}{d\tau}$$

Outside of a spherically symmetric source, the Schwarzschild solution for A and B is given in Eqs. (7.88).

---

[2] Compare Eq. (5.44). Parallel transport of a four-vector around this closed loop will yield a non-zero rotation.

Solution to Problem 7.3

The metric in spherical coordinates $q^\mu = (r, \theta, \phi, ct)$ is defined through the interval in Eq. (7.47)

$$(ds)^2 = A(r)(dr)^2 + r^2(d\theta)^2 + r^2 \sin^2 \theta (d\phi)^2 - B(r)(cdt)^2$$

The affine connection is calculated in the text and summarized in Table 7.1, which for convenience we reproduce below.

Table 7.1   SUMMARY —Affine connection for spherically symmetric source with coordinates $(q^1, \cdots, q^4) = (r, \theta, \phi, ct)$ and the metric of Eqs. (7.48). It is symmetric in its lower two indices so that $\Gamma^\lambda_{\mu\nu} = \Gamma^\lambda_{\nu\mu}$. Here $A' = dA(r)/dr$ and $B' = dB(r)/dr$.

$\Gamma^r_{rr} = \dfrac{A'}{2A}$	$\Gamma^r_{\phi\phi} = -\dfrac{r \sin^2 \theta}{A}$	$\Gamma^r_{\theta\theta} = -\dfrac{r}{A}$
$\Gamma^r_{44} = \dfrac{B'}{2A}$	$\Gamma^\theta_{\phi\phi} = -\sin\theta\cos\theta$	$\Gamma^\theta_{r\theta} = \dfrac{1}{r}$
$\Gamma^\phi_{r\phi} = \dfrac{1}{r}$	$\Gamma^\phi_{\theta\phi} = \dfrac{\cos\theta}{\sin\theta}$	$\Gamma^4_{r4} = \dfrac{B'}{2B}$

All others vanish

The geodesic Eq. (7.8) is then

$$\frac{d^2 q^\mu}{d\tau^2} + \Gamma^\mu_{\lambda\sigma} \frac{dq^\lambda}{d\tau} \frac{dq^\sigma}{d\tau} = 0 \qquad ; \text{ geodesics}$$

$$\mu = 1, \cdots, 4$$

The conversion from proper time to time $(\tau \to t)$ follows from the interval

$$(ds)^2 \equiv -(cd\tau)^2$$

Thus

$$\left(\frac{d\tau}{dt}\right)^2 = B(r) - \frac{1}{c^2}\left[A(r)(\dot{r})^2 + r^2(\dot{\theta})^2 + r^2 \sin^2 \theta (\dot{\phi})^2\right]$$

Hence

$$\frac{dt}{d\tau} = \left\{ B - \frac{1}{c^2}\left[(\dot{r})^2 A + (r\dot{\theta})^2 + (r\dot{\phi}\sin\theta)^2\right]\right\}^{-1/2} \equiv \rho$$

where a dot now indicates a derivative with respect to time. The geodesic equation is then re-written as

$$\frac{1}{\rho}\frac{d}{dt}(\rho \dot{q}^\mu) + \Gamma^\mu_{\lambda\sigma} \dot{q}^\lambda \dot{q}^\sigma = 0$$

We can now just start reading off the results with $q^\mu = (r, \theta, \phi, ct)$, and repeated Greek indices summed from 1 to 4. For the first component

$$\frac{1}{\rho}\frac{d}{dt}(\rho\dot{r}) + \left[\frac{A'}{2A}(\dot{r})^2 - \frac{r}{A}(\dot{\theta})^2 - \frac{r\sin^2\theta}{A}(\dot{\phi})^2 + \frac{B'}{2A}c^2\right] = 0$$

For the second component[3]

$$\frac{1}{\rho}\frac{d}{dt}(\rho\dot{\theta}) + \left[\frac{2}{r}(\dot{r}\dot{\theta}) - \sin\theta\cos\theta\,(\dot{\phi})^2\right] = 0$$

For the third component

$$\frac{1}{\rho}\frac{d}{dt}(\rho\dot{\phi}) + \left[\frac{2}{r}(\dot{r}\dot{\phi}) + 2\frac{\cos\theta}{\sin\theta}(\dot{\theta}\dot{\phi})\right] = 0$$

and for the fourth component

$$\frac{1}{\rho}\frac{d}{dt}(\rho c) + \left[\frac{B'}{B}(c\dot{r})\right] = 0$$

These are the stated results.

Problem 7.4 (a) Take the newtonian limit of the equations in Prob. 7.3 by letting $c^2 \to \infty$.

(b) The lagrangian in spherical coordinates for non-relativistic motion outside a spherically symmetric gravitational source is

$$L = T - V = \frac{m}{2}\vec{v}^2 + \frac{mMG}{r}$$

$$= \frac{m}{2}\left[(\dot{r})^2 + (r\dot{\theta})^2 + (r\dot{\phi}\sin\theta)^2\right] + \frac{mMG}{r}$$

Write Lagrange's equations for (r, θ, ϕ) and compare with the results for the first three equations in part (a).

(c) What is the content of the fourth equation in (a)?

Solution to Problem 7.4

(a) The Schwarzschild solution for $[A(r), B(r)]$ is given in Eqs. (7.88)

$$A(r) = \frac{1}{1 - 2MG/c^2 r} \qquad ; \; B(r) = 1 - \frac{2MG}{c^2 r}$$

---

[3]Recall $\Gamma^\theta_{\theta r} = \Gamma^\theta_{r\theta}$, and similarly for the others.

We use this as a guide in taking the $c^2 \to \infty$ limit of the equations in Prob. 7.3

$$A \to 1 \quad ; \ B \to 1 \quad ; \ B' \to O(1/c^2) \quad ; \ c^2 \to \infty$$

The first three equations then become[4]

$$\frac{d^2r}{dt^2} - r(\dot{\theta})^2 - r\sin^2\theta \, (\dot{\phi})^2 + \frac{B'}{2}c^2 = 0 \qquad ; \ c^2 \to \infty$$

$$\frac{d^2\theta}{dt^2} + \frac{2}{r}(\dot{r}\dot{\theta}) - \sin\theta\cos\theta \, (\dot{\phi})^2 = 0$$

$$\frac{d^2\phi}{dt^2} + \frac{2}{r}(\dot{r}\dot{\phi}) + 2\frac{\cos\theta}{\sin\theta}(\dot{\theta}\dot{\phi}) = 0$$

(b) Given the lagrangian $L(\dot{r}, \dot{\theta}, \dot{\phi}; \ r, \theta.\phi)$

$$L = \frac{m}{2}\left[(\dot{r})^2 + (r\dot{\theta})^2 + (r\dot{\phi}\sin\theta)^2\right] + \frac{mMG}{r}$$

Lagrange's equations follow as

$$\frac{d}{dt}\left[\frac{\partial L}{\partial(\partial L/\partial \dot{q}^i)}\right] - \frac{\partial L}{\partial q^i} = 0 \qquad ; \ q^i = (r, \theta.\phi)$$

Lagrange's equations then read

$$\frac{d^2r}{dt^2} - r(\dot{\theta})^2 - r\sin^2\theta \, (\dot{\phi})^2 + \frac{MG}{r^2} = 0 \qquad ; \ r\text{-eqn}$$

$$\frac{d}{dt}(r^2\,\dot{\theta}) - \sin\theta\cos\theta \, (r\dot{\phi})^2 = 0 \qquad ; \ \theta\text{-eqn}$$

$$\frac{d}{dt}(r^2\sin^2\theta\,\dot{\phi}) = 0 \qquad ; \ \phi\text{-eqn}$$

Evaluation of the time derivatives in the last two equations gives

$$\frac{d^2r}{dt^2} - r(\dot{\theta})^2 - r\sin^2\theta \, (\dot{\phi})^2 + \frac{MG}{r^2} = 0 \qquad ; \ r\text{-eqn}$$

$$\frac{d^2\theta}{dt^2} + \frac{2}{r}(\dot{r}\dot{\theta}) - \sin\theta\cos\theta \, (\dot{\phi})^2 = 0 \qquad ; \ \theta\text{-eqn}$$

$$\frac{d^2\phi}{dt^2} + \frac{2}{r}(\dot{r}\dot{\phi}) + 2\frac{\cos\theta}{\sin\theta}(\dot{\theta}\dot{\phi}) = 0 \qquad ; \ \phi\text{-eqn}$$

Provided we identify

$$B' = \frac{2MG}{c^2r^2}$$

---

[4]See part (c) for the fourth equation.

as suggested by the Schwarzschild solution, these now reproduce the geodesic equations in the metric in Eq. (7.47) in the $c^2 \to \infty$ limit.

(c) Although quite generally providing a valuable expression for the time development of ρ, in this limit the fourth geodesic equation just reduces to $\dot{\rho} = 0$.

Problem 7.5 (a) Look for a solution to the equations in Prob. 7.3 with constant $(r = a, \theta = \pi/2, \dot{\phi} = \omega)$ corresponding to circular motion in the Schwarzschild metric in Eqs. (7.88). Show the angular frequency is given by[5]

$$\omega^2 = \frac{MG}{a^3}$$

Compare with the newtonian result.

(b) Show the solution with constant $(r = a, \dot{\theta} = \omega, \phi = \phi_0)$ gives the same result as in (a).

Solution to Problem 7.5

(a) Look for a solution to the equations in Prob. 7.3 with constant

$$r = a \qquad ; \theta = \frac{\pi}{2} \qquad ; \dot{\phi} = \omega \qquad ; \text{constants}$$

This implies

$$\dot{r} = \dot{\theta} = 0$$

Only the radial equation then gives the non-zero result

$$a\omega^2 = \frac{B'}{2}c^2$$

With the Schwarzschild metric, one has (see Prob. 7.4)

$$B' = \frac{2MG}{c^2 a^2}$$

Hence for circular motion in the (x, y) plane, in the Schwarzschild metric, the angular frequency is

$$\omega^2 = \frac{MG}{a^3}$$

Note that no approximations have been made. The newtonian limit in Prob. 7.4 gives precisely the same result for ω^2.

---

[5]The period is $p = 2\pi/\omega$.

(b) Suppose one similarly looks for circular motion with

$$r = a \qquad ; \dot{\theta} = \omega \qquad ; \phi = \phi_0 \qquad ; \text{constants}$$

This implies

$$\dot{r} = \dot{\phi} = 0$$

Again, only the radial equation then gives the non-zero result[6]

$$a\omega^2 = \frac{B'}{2}c^2$$

Then, just as in part (a)

$$\omega^2 = \frac{MG}{a^3}$$

This is for circular motion, in the plane $\phi = \phi_0$, in the Schwarzschild metric.

Problem 7.6 (a) Look for a solution to the equations in Prob. 7.3 with constant (θ, ϕ) corresponding to purely radial motion. Show the equation of motion for the radial coordinate can be reduced to the form

$$\ddot{r} + \dot{r}^2 \left(\frac{A'}{2A} - \frac{B'}{B} \right) + \frac{B'}{2A}c^2 = 0$$

(b) Use the Schwarzschild solution of Eqs. (7.88), expand through first order in $1/c^2$, and show that the result in part (a) becomes

$$\ddot{r} + \frac{MG}{r^2} \left[1 - \frac{2MG}{c^2 r} - 3\frac{\dot{r}^2}{c^2} \right] = 0$$

Solution to Problem 7.6

(a) Assume purely radial motion, with

$$\dot{\theta} = \dot{\phi} = 0$$

The remaining non-zero first and fourth geodesic equations in Prob. 7.3 then become

$$\frac{d^2 r}{dt^2} + \frac{\dot{r}}{\rho} \frac{d\rho}{dt} + \frac{A'}{2A}\dot{r}^2 + \frac{B'}{2A}c^2 = 0$$

$$\frac{1}{\rho}\frac{d\rho}{dt} = -\dot{r}\frac{B'}{B}$$

---

[6] As in chap. 8, once the motion is confined to the plane with $\phi = \phi_0$, the generalized coordinate θ can take all values.

These combine to give

$$\ddot{r} + \left(\frac{A'}{2A} - \frac{B'}{B} \right) \dot{r}^2 + \frac{B'}{2A} c^2 = 0$$

This is the stated answer for radial motion in the metric in Eq. (7.47).[7]

(b) The Schwarzschild solutions for $[A(r), B(r)]$ are given in Eqs. (7.88)

$$A(r) = \frac{1}{1 - 2MG/c^2 r} \qquad ; \ B(r) = 1 - \frac{2MG}{c^2 r}$$

An expansion up through $O(1/c^2)$ then gives

$$\frac{A'}{2A} = -\frac{MG}{c^2 r^2}$$

$$\frac{B'}{B} = \frac{2MG}{c^2 r^2}$$

$$\frac{B'}{2A} c^2 = \frac{MG}{r^2} \left[1 - \frac{2MG}{c^2 r} \right]$$

Hence for radial motion in the Schwarzschild metric, through $O(1/c^2)$,

$$\ddot{r} + \frac{MG}{r^2} \left[1 - \frac{2MG}{c^2 r} - 3\frac{\dot{r}^2}{c^2} \right] = 0 \qquad ; \ \text{through O}(1/c^2)$$

Note that in the newtonian limit of Prob. 7.4, this become the familiar expression

$$\ddot{r} = -\frac{MG}{r^2} \qquad ; \ \text{newtonian limit}$$

Problem 7.7 (a) Convert the equations in Prob. 7.3 in the Schwarzschild metric to dimensionless form using R_s as the unit of length and R_s/c as the unit of time.

(b) Write, or obtain, a program to solve the resulting four, coupled, second-order, nonlinear differential equations.

(c) What is the role of the fourth equation?

(d) Compute some representative particle trajectories. Compare with the newtonian limit.

Solution to Problem 7.7

We will do part (d) first and then build on that.

---

[7]This complicated non-linear differential equation, derived without approximation, is now best handled with numerical integration.

(d) In a plane with constant $\phi = \phi_0$ (we take $\phi_0 = 0$), Newton's laws in Prob. 7.4 read

$$\frac{d^2r}{dt^2} = \frac{l^2}{m^2r^3} - \frac{MG}{r^2} \qquad ; l = mr^2\frac{d\theta}{dt}$$

$$\frac{dl}{dt} = 0$$

The dimensionless variables to be used are

$$u = \frac{r}{R_s} \qquad ; \sigma = \frac{ct}{R_s} \qquad ; L^2 = \frac{l^2}{(mcR_s)^2}$$

where $R_s = 2MG/c^2$ is the Schwarzschild radius. Then

$$\frac{d^2u}{d\sigma^2} = \frac{L^2}{u^3} - \frac{1}{2u^2} \qquad ; L^2 = u^4\left(\frac{d\theta}{d\sigma}\right)^2$$

$$\frac{dL}{d\sigma} = 0$$

Introduce the column vector[8]

$$\underline{Z}_n = \begin{bmatrix} u_n \\ (du/d\sigma)_n \\ \theta_n \end{bmatrix}$$

Newton's laws can then be recast as the following dimensionless, finite-difference, matrix equation

$$\underline{Z}_{n+1} = \underline{Z}_n + \Delta \begin{bmatrix} (Z_n)_2 \\ L^2/(Z_n)_1^3 - 1/2(Z_n)_1^2 \\ -L/(Z_n)_1^2 \end{bmatrix}$$

This equation can be iterated with Mathcad 7, given a starting value $\underline{Z}_1$.[9]

The initial conditions for the representative scattering trajectory in Fig. 7.1 are

$$\underline{Z}_1 = \begin{bmatrix} 15 \\ -0.5 \\ 3 \end{bmatrix} \qquad ; L = 1.10$$

---

[8]Readers should note that (u, θ, σ) are now simply the dimensionless coordinates (r, θ, t); to avoid any confusion, we give them new names. Here the components of $\underline{Z}_n$ are denoted $(Z_n)_i$ with $i = (1, 2, 3)$.

[9]We used the Mathcad matrix notation and $\Delta = 10^{-3}$. Note the sign of the integration constant in the third relation; θ is *decreasing* along the trajectory (see Fig. 7.8 in the text with $\phi = 0$).

Here $L = 1.10$ was adjusted to give a vertical velocity at $\sigma = 0$. The particle moves up along the trajectory.

NEWTONIAN TRAJECTORY

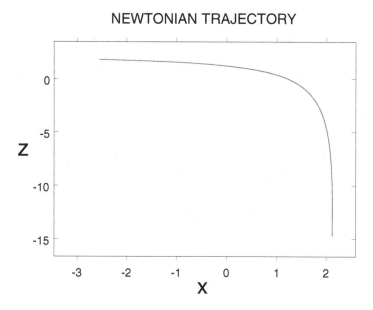

Fig. 7.1 Newtonian trajectory in the central gravitational potential $V(r) = -mMG/r$ with initial conditions $u_0 = 15$, $(du/d\sigma)_0 = -0.5$, $\theta_0 = 3$ and $L = 1.10$ adjusted to give an initial vertical velocity. The particle moves up along the trajectory. Here (x, z) are the familiar (dimensionless) cartesian coordinates in the plane with $\phi = 0$, and the source sits at the origin $(0, 0)$. Note the distinct scales.

(a) The geodesic equations in the Schwarzschild metric are derived from a lagrangian in chapter 8. The radial equation of motion, through order $1/c^2$, is given in Eqs. (8.32)

$$\frac{d}{dt}\left\{\dot{r}\left[1 + \frac{\dot{r}^2}{2c^2} + \frac{r^2}{2c^2}\left(\frac{l}{mr^2}\right)^2 + \frac{3MG}{rc^2}\right]\right\} =$$

$$\frac{l^2}{m^2r^3}\left[1 - \frac{\dot{r}^2}{2c^2} - \frac{r^2}{2c^2}\left(\frac{l}{mr^2}\right)^2 - \frac{3MG}{2rc^2}\right] - \frac{MG}{r^2}\left[1 + \frac{3\dot{r}^2}{2c^2} + \frac{MG}{rc^2}\right]$$

The corresponding angular equation of motion is given through the same order in Eqs. (8.23)–(8.25)

$$l = mr^2\dot\theta \left(1 + \frac{\dot r^2}{2c^2} + \frac{r^2\dot\theta^2}{2c^2} + \frac{MG}{c^2 r}\right)$$

$$\frac{dl}{dt} = 0$$

The definition of l is inverted through this order in Eq. (8.30)

$$\dot\theta = \frac{l}{mr^2}\left[1 - \frac{\dot r^2}{2c^2} - \frac{MG}{c^2 r} - \frac{r^2}{2c^2}\left(\frac{l}{mr^2}\right)^2\right]$$

The $c^2 \to \infty$ limit of these equations give the newtonian expressions studied in part (d) above. Here we examine the effect of these $O(1/c^2)$ terms on the trajectory.

The new column vector in the dimensionless variables of part (d) becomes

$$\underline{Z}_n = \begin{bmatrix} u_n \\ [F(du/d\sigma)]_n \\ \theta_n \end{bmatrix} \quad ; \quad F\left(u, \frac{du}{d\sigma}\right) = \left[1 + \frac{1}{2}\left(\frac{du}{d\sigma}\right)^2 + \frac{1}{2}\frac{L^2}{u^2} + \frac{3}{2u}\right]$$

with an initial value $\underline{Z}_1$. The new finite-difference equations then take the form

$$\underline{Z}_{n+1} = \underline{Z}_n + \Delta \begin{bmatrix} (du/d\sigma)_n \\ (L^2/u_n^3)\,G_n - (1/2u_n^2)\,H_n \\ -(L/u_n^2)\,I_n \end{bmatrix}$$

where

$$G\left(u, \frac{du}{d\sigma}\right) = \left[1 - \frac{1}{2}\left(\frac{du}{d\sigma}\right)^2 - \frac{1}{2}\frac{L^2}{u^2} - \frac{3}{4u}\right]$$

$$H\left(u, \frac{du}{d\sigma}\right) = \left[1 + \frac{3}{2}\left(\frac{du}{d\sigma}\right)^2 + \frac{1}{2u}\right]$$

$$I\left(u, \frac{du}{d\sigma}\right) = \left[1 - \frac{1}{2}\left(\frac{du}{d\sigma}\right)^2 - \frac{1}{2}\frac{L^2}{u^2} - \frac{1}{2u}\right]$$

At each point, the derivative $(du/d\sigma)_n$ must be found from a given value of the quantity $[F(du/d\sigma)]_n$

$$\left(\frac{du}{d\sigma}\right)_n \left[1 + \frac{1}{2}\left(\frac{du}{d\sigma}\right)_n^2 + \frac{1}{2}\frac{L^2}{u_n^2} + \frac{3}{2u_n}\right] = \left[F\left(\frac{du}{d\sigma}\right)\right]_n$$

To be consistent with the $1/c^2$ expansion, the correction terms must remain small.

The initial conditions for the representative geodesic in Fig. 7.2 are

$$\underline{Z}_1 = \begin{bmatrix} 10 \\ -0.22 \\ 2.5 \end{bmatrix} \qquad ; L = 1.92$$

Here $L = 1.92$ was adjusted to give a vertical velocity at $\sigma = 0$.[10] The particle moves up along the trajectory. This geodesic is compared with

SCHWARZSCHILD GEODESIC TO ORDER 1/c²

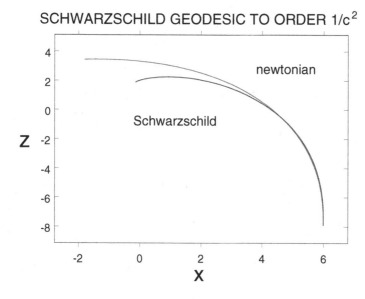

Fig. 7.2 A representative geodesic in the Schwarzschild metric that retains $1/c^2$ corrections to the newtonian orbit [see part (d) and chap. 8]. The initial conditions are $u_0 = 10$, $[F(du/d\sigma)]_0 = -0.22$, $\theta_0 = 2.5$; and $L = 1.92$ was adjusted to give an initial vertical velocity. The particle moves up along the trajectory. This is compared to the newtonian orbit with an adjusted $L = 1.72$.

---

[10]We use $\Delta = .0015$ and a Mathcad solveblock to find $(du/d\sigma)_n$. We also monitor the corrections along the trajectory to verify that they remain small.

the corresponding newtonian orbit, which neglects all the correction terms and has an adjusted value of $L = 1.72$. Note that the geodesic contains all the corrections arising from both special and general relativity.

To obtain some insight, we can simply look at the overall *sign* of the correction appearing in the various terms in the finite-difference equation:[11]

- Through H, the gravitational force is *increased*;
- Through G, the angular momentum barrier is *decreased*;
- Through I, the rate of decrease of the angle is *lessened*;
- Through F, the magnitude of $du/d\sigma$ is *lessened*.

All of these effects contribute to the falling in of the Schwarzschild geodesic from the newtonian trajectory in Fig. 7.2.

(b,c) We are now in a position to tackle the full geodesic. With constant ϕ, and the use of the fourth equation to eliminate $\dot{\rho}/\rho$, the equations in Prob. 7.3 become

$$\ddot{r} + \left[\left(\frac{A'}{2A} - \frac{B'}{B}\right)(\dot{r})^2 - \frac{r}{A}(\dot{\theta})^2 + \frac{B'}{2A}c^2\right] = 0$$

$$\ddot{\theta} + \left(\frac{2}{r} - \frac{B'}{B}\right)(\dot{r}\dot{\theta}) = 0$$

Substitution of the Schwarzschild solution in Eqs. (7.88)

$$A(r) = \frac{1}{1 - 2MG/c^2 r} \qquad ; \ B(r) = 1 - \frac{2MG}{c^2 r}$$

leads to the relations

$$\ddot{r} + \left[\frac{(-3MG/c^2 r^2)}{(1 - 2MG/c^2 r)}(\dot{r})^2 - r\left(1 - \frac{2MG}{c^2 r}\right)(\dot{\theta})^2 + \frac{MG}{r^2}\left(1 - \frac{2MG}{c^2 r}\right)\right] = 0$$

$$\ddot{\theta} + \left[\left(\frac{2}{r}\right)\frac{(1 - 3MG/c^2 r)}{(1 - 2MG/c^2 r)}\right](\dot{r}\dot{\theta}) = 0$$

Written in terms of the dimensionless variables in part (d), these equations take the form

$$\frac{d^2 u}{d\sigma^2} + \frac{1}{2u^3}(u - 1) - \frac{3}{2u(u-1)}\left(\frac{du}{d\sigma}\right)^2 - (u - 1)\left(\frac{d\theta}{d\sigma}\right)^2 = 0$$

$$\frac{d^2\theta}{d\sigma^2} + \frac{2}{u}\frac{(u - 3/2)}{(u - 1)}\left(\frac{du}{d\sigma}\right)\left(\frac{d\theta}{d\sigma}\right) = 0$$

---

[11]It is amusing that all of the corrections in each of the following expressions have the same sign.

These coupled, non-linear, second-order differential equations can again be solved with the iteration of a dimensionless, finite-difference, matrix equation based on the four-component column vector

$$\underline{Z} = \begin{bmatrix} u \\ du/d\sigma \\ \theta \\ d\theta/d\sigma \end{bmatrix}$$

The finite-difference equation reads

$$\underline{Z}_{n+1} = \underline{Z}_n + \Delta \times$$

$$\begin{bmatrix} (du/d\sigma)_n \\ -(u_n - 1)/2u_n^3 + [3/2u_n(u_n - 1)](du/d\sigma)_n^2 + (u_n - 1)(d\theta/d\sigma)_n^2 \\ (d\theta/d\sigma)_n \\ -[2(u_n - 3/2)/u_n(u_n - 1)](du/d\sigma)_n(d\theta/d\sigma)_n \end{bmatrix}$$

The newtonian limit is[12]

$$\underline{Z}_{n+1} = \underline{Z}_n + \Delta \begin{bmatrix} (\overset{\bullet}{du}/d\sigma)_n \\ -1/2u_n^2 + u_n(d\theta/d\sigma)_n^2 \\ (d\theta/d\sigma)_n \\ -(2/u_n)(du/d\sigma)_n(d\theta/d\sigma)_n \end{bmatrix} \quad ; \text{ newtonian}$$

The finite difference equations were again iterated with Mathcad 7. One full geodesic is shown in Fig. 7.3. The initial conditions are

$$\underline{Z}_1 = \begin{bmatrix} 10 \\ -0.23 \\ 2.5 \\ -1.80 \times 10^{-2} \end{bmatrix}$$

The quantity $(d\theta/d\sigma)_0$ was varied until the initial trajectory is vertical. The particle moves up along the trajectory. This is compared to the newtonian orbit with the same initial conditions.[13] Note that Fig. 7.3 is what the record keeper sees on his or her screen.

The full result in Fig. 7.3 should be compared to the one in Fig. 7.2, which retains only the $1/c^2$ corrections to the newtonian orbit, but which

---

[12]We have factored c^2/R_s from the radial equation and c^2/R_s^2 from the angular equation, leaving $u = O(c^2)$ and $\sigma = O(c^3)$ as $c \to \infty$.

[13]Here $\Delta = 5 \times 10^{-4}$, and the result is stable to decreasing in Δ while maintaining the same total range.

FULL GEODESIC IN SCHWARZSCHILD METRIC

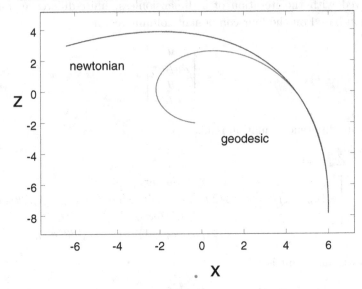

Fig. 7.3   A full geodesic in the Schwarzschild metric. The initial conditions are $u_0 = 10$, $(du/d\sigma)_0 = -0.23$, $\theta_0 = 2.5$, and $(d\theta/d\sigma)_0 = -1.80 \times 10^{-2}$ was varied until the initial trajectory is vertical. The particle moves up along the trajectory. This is compared to the newtonian orbit with the same initial conditions. Here $\Delta = 5 \times 10^{-4}$.

nicely anticipates the full behavior. It is fascinating that far-enough along the orbit, the behavior of the two curves in Fig. 7.3 is qualitatively different.

Readers are urged to investigate other geodesics. Just to whet the appetite, Fig. 7.4 shows an extension of Fig. 7.3 to where the radial coordinate gets down to the Schwarzschild radius.[14]

Problem 7.8 The interior of a satellite orbiting outside of a spherically symmetric gravitational source also forms a local freely falling frame (LF^3) since the inertial centrifugal force just balances the gravitational attraction. Suppose one has a clock on a satellite which is in a circular orbit at a distance of $r = 10^8 R_s$ from the center of force, and which orbits it for 2000 days. What is the time difference (in sec) coming from general relativity between that clock and one at rest in the inertial laboratory frame very far away from the source. Work through $O(1/c^2)$.

---

[14]Readers can then instruct the author as to the role of energy conservation in these geodesics! Recall from Prob. 7.5 that there are bound circular geodesics in the Schwarzschild metric, and from chap. 8, at least through $O(1/c^2)$, these orbits are stable.

FULL GEODESIC IN SCHWARZSCHILD METRIC

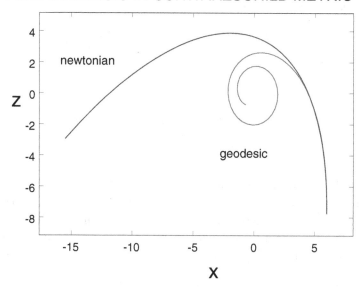

Fig. 7.4 Extension of the range in Fig. 7.3 to where the radial coordinate in the geodesic gets down to the Schwarzschild radius. The two curves have the same duration in σ.

Solution to Problem 7.8

The Schwarzschild metric is given in Eq. (7.89)

$$(ds)^2 = \frac{(dr)^2}{1 - 2MG/c^2r} + r^2(d\theta)^2 + r^2\sin^2\theta(d\phi)^2 - \left(1 - \frac{2MG}{c^2r}\right)(cdt)^2$$

In the LF^3 of the clock, the interval is just the proper time

$$(ds)^2 = -(cd\tau)^2$$

With the circular motion of Prob. 7.5, one has

$$r = a \qquad ; \ \theta = \frac{\pi}{2} \qquad ; \ \dot{\phi} = \omega \qquad\qquad ; \ \text{constants}$$

Hence

$$-(cd\tau)^2 = (a\omega)^2 dt^2 - \left(1 - \frac{2MG}{c^2a}\right)(cdt)^2$$

$$dt = \left[1 - \frac{(a\omega)^2}{c^2} - \frac{2MG}{c^2a}\right]^{-1/2} d\tau$$

This illustrates the time dilation of both special relativity and the gravitational field.[15] Use (see Prob. 7.5)

$$\omega^2 = \frac{MG}{a^3} \qquad ; \ R_s = \frac{2MG}{c^2}$$

Then

$$dt = \left(1 - \frac{3R_s}{2a}\right)^{-1/2} d\tau$$

$$\approx \left(1 + \frac{3R_s}{4a}\right) d\tau$$

where we have expanded through $O(1/c^2)$.

Thus the difference in general relativity between a clock in circular orbit at a radius a in the Schwarzschild metric, where the interval in the record keeper's laboratory time coordinate is Δt, and one at rest in the LF^3 of the satellite, where the clock registers the proper time interval $\Delta\tau$, is[16]

$$\Delta t - \Delta\tau = \frac{3R_s}{4a}\Delta\tau$$

Now use

$$a = 10^8 \, R_s$$
$$\Delta\tau = 2000 \, \text{days} = 1.73 \times 10^8 \, \text{sec}$$

This gives

$$\Delta t - \Delta\tau = 1.30 \, \text{sec}$$

Problem 7.9 A standard meter stick, very small on that scale, lies in the surface in Fig. 7.9 in the text and is oriented in the radial direction. It is then slid inward toward the symmetry axis. The location of its two ends at a given instant in t are reported to the record keeper, who plots the two points in Fig. 7.10 in the text.

(a) What does the record keeper actually see?

(b) How is the record keeper able to keep track of what is happening physically?

---

[15]Note $a\omega = v$

[16]There is a *time dilation* for the moving clock in general relativity. The proper time is identical to what one measures in a frame that is at rest far away from the source.

Solution to Problem 7.9

(a) The Schwarzschild metric is given in Eq. (7.89)

$$(ds)^2 = \frac{(dr)^2}{1 - 2MG/c^2 r} + r^2(d\theta)^2 + r^2 \sin^2 \theta (d\phi)^2 - \left(1 - \frac{2MG}{c^2 r}\right)(cdt)^2$$

At constant coordinate time t. and given angles (θ, ϕ), one has

$$dt = d\theta = d\phi = 0 \qquad ; (t, \theta, \phi) \text{ constant}$$

Now suppose one has a (short) meter stick of length l_0 oriented in the radial direction. Then in the LF^3 the interval is just

$$(ds)^2 = l_0^2$$

If the length of the meter stick in the lab frame is denoted by $dr = l$, then the length of the meter stick in the Schwarzschild metric, at a given coordinate time t, satisfies[17]

$$l_0^2 = \frac{1}{1 - 2MG/c^2 r} l^2$$

This illustrates the *length contraction* in general relativity

$$l^2 = \left(1 - \frac{R_s}{r}\right) l_0^2 \qquad ; R_s = \frac{2MG}{c^2}$$

If the length l^2 is reported to the record keeper as a function of the radial coordinate distance expressed as R_s/r, then the record keeper sees what is in Fig. 7.5. At $R_s/r = 0$, far away from the source, the length is $l^2 = l_0^2$. At the Schwarzschild radius $R_s/r = 1$, the length has disappeared and $l^2 = 0$!

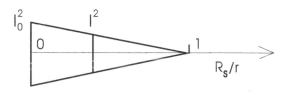

Fig. 7.5 Length l of a meter stick of proper length l_0 oriented in the radial direction in the Schwarzschild metric, at a given coordinate time t and radial coordinate r, as seen by the record keeper. Here l^2 is plotted as a funtion of R_s/r.

---

[17]The general expression with the metric for a spherically symmetric source in Eq. (7.47) is $l_0^2 = A(r)l^2$.

(b) Figure 7.9 in the text helps to clarify what is going on. As the meter stick, oriented in the radial direction, is slid along the surface toward the symmetry axis, the radial distances r to the two ends of the meter stick approach each other. Thus, although the true length of the meter stick remains l_0, the coordinate length l that the record keeper sees *does* shrink. At the Schwarzschild radius, the meter stick is vertical in Fig.7.9 in the text, and its coordinate length l, at a given coordinate time t indeed vanishes.

It is the *metric* that allows the record keeper to arrive at a proper physical interpretation of his or her coordinate results.[18]

---

[18]An important lesson is reinforced here. Coordinate results by themselves are physically meaningless without the metric.

Chapter 8

Precession of Perihelion

Problem 8.1 This problem constructs the hamiltonian corresponding to the lagrangian in Eq. (8.22).[1]

(a) Show through $O(1/c^2)$ that

$$\dot{\theta} = \frac{p_\theta}{mr^2}\left(1 - \frac{v^2}{2c^2} + \frac{\Phi}{c^2}\right)$$

$$\dot{r} = \frac{p_r}{m}\left(1 - \frac{v^2}{2c^2} + \frac{\Phi}{c^2}\right) + \frac{2p_r}{m}\frac{\Phi}{c^2}$$

(b) Show through $O(1/c^2)$ that the hamiltonian $H = p_r\dot{r} + p_\theta\dot{\theta} - L$ is given by

$$H = T_0\left(1 - \frac{T_0}{2mc^2}\right) + V_0\left(1 + \frac{T_0}{mc^2} - \frac{V_0}{2mc^2} + \frac{1}{mc^2}\frac{p_r^2}{m}\right) + mc^2$$

Here

$$T_0 \equiv \frac{1}{2m}\left(p_r^2 + \frac{p_\theta^2}{r^2}\right) \qquad ; V_0 \equiv -\frac{mMG}{r}$$

(c) The lagrangian in Eq. (8.22) has no explicit dependence on time so that $\partial L/\partial t = 0$, and it is cyclic in θ. Use these facts to show that the result in (b) produces a first integral of the radial equation of motion in the form

$$H(p_r, r) = \text{constant}$$

---

[1] *Hint:* It is simplest to start with Eqs. (8.14) and (8.21).

Solution to Problem 8.1

(a) The lagrangian $L(r, \theta; \dot{r}, \dot{\theta})$ in Eq. (8.22) for a particle of rest mass m in the Schwarzschild metric is

$$L \doteq \frac{m}{2}(\dot{r}^2 + r^2\dot{\theta}^2) + \frac{m}{8c^2}(\dot{r}^4 + 2\dot{r}^2r^2\dot{\theta}^2 + r^4\dot{\theta}^4) +$$

$$\frac{mMG}{r} + \frac{mMG}{r}\frac{\dot{r}^2}{c^2} + \frac{mMG\,(\dot{r}^2 + r^2\dot{\theta}^2)}{r}\frac{}{2c^2} + \frac{m}{2c^2}\left(\frac{MG}{r}\right)^2$$

The symbol $\doteq$ indicates that the constant term $-mc^2$ has been dropped. This expression is then correct through $O(1/c^2)$. In the following, we will make use of the expression for the gravitational potential Φ, and for v^2 in Eq. (8.21),

$$\Phi(r) = -\frac{MG}{r} \qquad ; v^2 = \dot{r}^2 + r^2\dot{\theta}^2$$

From Eqs. (8.24)–(8.25), the canonical momentum p_θ is

$$p_\theta = \frac{\partial L}{\partial \dot{\theta}} = mr^2\dot{\theta}\left(1 + \frac{\dot{r}^2}{2c^2} + \frac{r^2\dot{\theta}^2}{2c^2} + \frac{MG}{c^2 r}\right)$$

$$= mr^2\dot{\theta}\left(1 + \frac{v^2}{2c^2} - \frac{\Phi}{c^2}\right)$$

This expression is inverted through $O(1/c^2)$ as

$$\dot{\theta} = \frac{p_\theta}{mr^2}\left(1 - \frac{v^2}{2c^2} + \frac{\Phi}{c^2}\right)$$

In a similar fashion, the canonical momentum p_r is

$$p_r = \frac{\partial L}{\partial \dot{r}} = m\dot{r}\left(1 + \frac{\dot{r}^2}{2c^2} + \frac{r^2\dot{\theta}^2}{2c^2} + \frac{3MG}{c^2 r}\right)$$

$$= m\dot{r}\left(1 + \frac{v^2}{2c^2} - \frac{3\Phi}{c^2}\right)$$

This expression is also inverted through $O(1/c^2)$ as

$$\dot{r} = \frac{p_r}{m}\left(1 - \frac{v^2}{2c^2} + \frac{\Phi}{c^2}\right) + \frac{2p_r}{m}\frac{\Phi}{c^2}$$

(b) The hamiltonian $H(p_r, p_\theta; r, \theta)$ is given by

$$H = p_r\dot{r} + p_\theta\dot{\theta} - L$$

This is evaluated up through $O(1/c^2)$ as

$$H = \frac{1}{m}\left(p_r^2 + \frac{p_\theta^2}{r^2}\right)\left(1 - \frac{v^2}{2c^2} + \frac{\Phi}{c^2}\right) + \frac{2p_r^2\Phi}{mc^2} -$$

$$\left\{\frac{1}{2m}\left(p_r^2 + \frac{p_\theta^2}{r^2}\right)\left(1 - \frac{v^2}{c^2} + \frac{2\Phi}{c^2}\right) + \frac{2p_r^2\Phi}{mc^2} + \frac{mv^4}{8c^2} + \right.$$

$$\left.\frac{mMG}{r}\left(1 + \frac{p_r^2}{m^2c^2} + \frac{v^2}{2c^2} + \frac{MG}{2rc^2}\right)\right\} + mc^2$$

where we have restored the rest mass mc^2. Now introduce

$$T_0 \equiv \frac{1}{2m}\left(p_r^2 + \frac{p_\theta^2}{r^2}\right) \qquad ; V_0 \equiv m\Phi = -\frac{mMG}{r}$$

In the correction terms of $O(1/c^2)$, one can use

$$v^2 = \dot{r}^2 + r^2\dot{\theta}^2 = \frac{1}{m^2}\left(p_r^2 + \frac{p_\theta^2}{r^2}\right) = \frac{2}{m}T_0$$

Hence

$$H = T_0\left(1 - \frac{T_0}{2mc^2}\right) + V_0\left(1 + \frac{T_0}{mc^2} - \frac{V_0}{2mc^2} + \frac{p_r^2}{m^2c^2}\right) + mc^2$$

This is the stated answer.

To the given order, this expression can be re-written after some algebra as

$$H = \frac{mc^2}{(1 - v^2/c^2)^{1/2}} \times$$

$$\left\{\left[1 + \frac{2MG}{c^2r}\left(\frac{\dot{r}^2}{c^2} + \frac{v^2}{2c^2}\right)\right] - \frac{MG}{c^2r}\left[1 + \frac{\dot{r}^2}{c^2} + \frac{1}{2c^2}\left(\frac{MG}{r}\right)\right]\right\}$$

If we employ the relativistic energy $\tilde{m}c^2 \equiv mc^2/(1 - v^2/c^2)^{1/2}$, and then work through $O(1/c^2)$ in the curly brackets, this becomes

$$H = \tilde{m}c^2 - \frac{\tilde{m}MG}{r} \qquad ; \tilde{m} \equiv \frac{m}{(1 - v^2/c^2)^{1/2}}$$

$$= \tilde{E}$$

This is the total energy $\tilde{E}$ of a particle of relativistic mass $\tilde{m}$ moving in the gravitational potential $\Phi = -MG/r$.[2] Through $O(1/c^4)$ in the curly

---

[2] Note that for $\Phi = 0$, this reduces to $H = \tilde{m}c^2$, which is the correct result for relativistic particle motion (see Prob. 6.2).

brackets, the strength of the potential and the relativistic energy are each increased with two general-relativistic corrections.

(c) Hamilton's equations of motion with a hamiltonian $H(p_i, q_i; t)$ and coordinates $q_i = (r, \theta)$ are[3]

$$\frac{dq_i}{dt} = \frac{\partial H}{\partial p_i} \qquad ; \frac{dp_i}{dt} = -\frac{\partial H}{\partial q_i} \qquad ; \frac{dH}{dt} = \frac{\partial H}{\partial t}$$

The coordinate θ is *cyclic* (it does not appear in H), and hence the angular momentum p_θ is a constant of the motion [see Eq. (8.23)]

$$p_\theta \equiv l = \text{constant} \qquad ; \text{angular momentum}$$

Substitute this in the above, so that

$$T_0 \equiv \frac{1}{2m}\left(p_r^2 + \frac{l^2}{r^2}\right) \qquad ; V_0 \equiv m\Phi = -\frac{mMG}{r}$$

The hamiltonian in part (a) has no explicit dependence on the time, and hence it is now *also* a constant of the motion

$$H(p_r, r) = \text{constant}$$

Several comments:

- This is an *important* result;
- It provides a rigorous constant of the motion through $O(1/c^2)$;
- In continuum mechanics, the energy density is obtained from the fourth component of the energy-momentum tensor, which agrees with the result obtained from the lagrangian density through the canonical procedure (see the *Aside* in Prob. 12.5);
- We will assume this also applies here, and we associate $H(p_r, r)$ with the total particle energy

$$H(p_r, r) = E_{\text{tot}} \qquad ; \text{total particle energy}$$

- Energy is then conserved along the particle trajectory;
- Far away from the source as $r \to \infty$, this energy reduces to the correct relativistic expression [recall part (b)]

$$E_{\text{tot}} = \frac{mc^2}{(1 - v^2/c^2)^{1/2}} \qquad ; r \to \infty$$

---

[3]See [Fetter and Walecka (2003)].

Problem 8.2 The newtonian limit $c^2 \to \infty$ in Prob. 8.1 (NRL) produces the familiar statement of conservation of energy

$$T_0 + V_0 = \frac{1}{2m}\left(p_r^2 + \frac{l^2}{r^2}\right) - \frac{mMG}{r} = E = \text{constant}$$

where $l = \text{constant}$, and $E = H - mc^2$. Define the effective potential by

$$\frac{1}{2m}p_r^2 + V_{\text{eff}}(r) = E$$

(a) Show that for circular orbits of radius a at the minimum of the effective potential

$$\frac{l^2}{m^2 a^3} = \frac{MG}{a^2} \qquad ; E_0 = -\frac{mMG}{2a}$$

(b) Assume $\delta E = E - E_0$ and $\delta r = r - a$ are small and expand about the minimum of the effective potential. Show the phase space orbit is an ellipse given by

$$\frac{p_r^2}{A^2} + \frac{(\delta r)^2}{B^2} = 1$$

$$A^2 = 2m\,\delta E \qquad ; B^2 = \frac{2a^3}{mMG}\delta E$$

(c) Show the area of the ellipse in (b) is $\mathcal{A} = 2\pi\,\delta E/\dot\theta_0$ where $\dot\theta_0$ is defined in Eq. (8.37).

Solution to Problem 8.2

(a) The non-relativistic limit (NRL) of the hamiltonian in Prob. 8.1, obtained for $c^2 \to \infty$, is

$$H - mc^2 = T_0 + V_0 = \frac{p_r^2}{2m} + V_{\text{eff}}(r) \qquad ; \text{NRL}$$

$$V_{\text{eff}}(r) = \frac{l^2}{2mr^2} - \frac{mMG}{r}$$

From Prob. 8.1(c) the hamiltonian, here the energy, is constant

$$H - mc^2 \equiv E = \text{constant}$$

The minimum of the effective potential occurs at

$$V'_{\text{eff}}(a) = -\frac{l^2}{ma^3} + \frac{mMG}{a^2} = 0$$

Hence, for a particle at rest in this minimum

$$\frac{l^2}{m^2 a^3} = \frac{MG}{a^2} \qquad ; E_0 = -\frac{mMG}{2a}$$

(b) Now assume that the particle is given a small transverse kick, so that l remains the same,[4] but there is now some small radial motion with $p_r^2 > 0$. Then

$$r = a + \delta r \qquad ; E = E_0 + \delta E$$

An expansion through second order in $\delta r/a$ then gives

$$E_0 + \delta E = \frac{p_r^2}{2m} + \frac{l^2}{2ma^2}\left[1 - 2\frac{\delta r}{a} + 3\frac{(\delta r)^2}{a^2} + \cdots\right]$$

$$- \frac{mMG}{a}\left[1 - \frac{\delta r}{a} + \frac{(\delta r)^2}{a^2} + \cdots\right]$$

$$= \frac{p_r^2}{2m} + E_0 + \frac{mMG(\delta r)^2}{2a^3}$$

which is rearranged to read

$$\frac{p_r^2}{A^2} + \frac{(\delta r)^2}{B^2} = 1$$

$$A^2 = 2m\,\delta E \qquad ; B^2 = \frac{2a^3}{mMG}\delta E$$

This describes an ellipse in the phase space $(p_r, \delta r)$, with semi-major axis A and semi-minor axis B.

(c) The area $\mathcal{A}$ of the ellipse is

$$\mathcal{A} = \pi AB = \pi \left[\frac{4a^3}{MG}\right]^{1/2} \delta E$$

From Prob. 7.5, the angular velocity for circular motion in this potential is[5]

$$\omega^2 = \frac{MG}{a^3}$$

Hence the area of the phase-space ellipse is

$$\mathcal{A} = \frac{2\pi\delta E}{\omega}$$

---

[4]Note $\vec{l} = \vec{r} \times \vec{p}$, and with $\delta\vec{p} \propto \vec{r}$, then $\delta\vec{l} = 0$.
[5]Compare Eq. (8.37).

Problem 8.3 Consider the radial Eq. (8.32):

(a) Show that purely radial motion is characterized by $l = 0$;

(b) Show the result obtained with $l = 0$ reproduces that in Prob. 7.6(b).

Solution to Problem 8.3

(a) The radial Eq. (8.32) derived from the lagrangian L in Eq. (8.22) and Prob. 8.1, for a particle of rest mass m in the Schwarzschild metric is

$$\frac{d}{dt}\left\{\dot{r}\left[1 + \frac{\dot{r}^2}{2c^2} + \frac{r^2}{2c^2}\left(\frac{l}{mr^2}\right)^2 + \frac{3MG}{rc^2}\right]\right\} =$$

$$\frac{l^2}{m^2 r^3}\left[1 - \frac{\dot{r}^2}{2c^2} - \frac{r^2}{2c^2}\left(\frac{l}{mr^2}\right)^2 - \frac{3MG}{2rc^2}\right] - \frac{MG}{r^2}\left[1 + \frac{3\dot{r}^2}{2c^2} + \frac{MG}{rc^2}\right]$$

The equation is exact through $O(1/c^2)$. The angular momentum l is given in Eq. (8.25)

$$l = mr^2\dot{\theta}\left(1 + \frac{\dot{r}^2}{2c^2} + \frac{r^2\dot{\theta}^2}{2c^2} + \frac{MG}{c^2 r}\right) \qquad ; \text{ angular momentum}$$

For purely radial motion one has constant θ, and $\dot{\theta} = 0$. Hence l vanishes[6]

$$l = 0 \qquad ; \text{ radial motion}$$

In this case, the radial equation reduces to

$$\frac{d}{dt}\left\{\dot{r}\left[1 + \frac{\dot{r}^2}{2c^2} + \frac{3MG}{rc^2}\right]\right\} = -\frac{MG}{r^2}\left[1 + \frac{3\dot{r}^2}{2c^2} + \frac{MG}{rc^2}\right]$$

(b) Evaluation of the time derivative in the above gives

$$\ddot{r}\left[1 + \frac{3\dot{r}^2}{2c^2} + \frac{3MG}{rc^2}\right] - \frac{3MG}{r^2}\frac{\dot{r}^2}{c^2} = -\frac{MG}{r^2}\left[1 + \frac{3\dot{r}^2}{2c^2} + \frac{MG}{rc^2}\right]$$

Thus, through $O(1/c^2)$,

$$\ddot{r} = -\frac{MG}{r^2}\left[1 - 3\frac{\dot{r}^2}{c^2} - \frac{2MG}{rc^2}\right] \qquad ; \text{ through } O(1/c^2)$$

This is identical to the result from the geodesic equations for radial motion in the Schwarzschild metric in Prob. 7.6(b)

$$\ddot{r} + \frac{MG}{r^2}\left[1 - \frac{2MG}{c^2 r} - 3\frac{\dot{r}^2}{c^2}\right] = 0 \qquad ; \text{ through } O(1/c^2)$$

---

[6]This corresponds to the notion that $\vec{l} = \vec{r} \times \vec{p}$ vanishes when $\vec{p}$ is along $\vec{r}$.

Problem 8.4 The reconciliation of the angular geodesic equations in Prob. 7.3 with the conservation of the angular momentum l of Eq. (8.25) in the lagrangian approach is, in fact, quite subtle. This problem leads the reader through that reconciliation. As in the text, assume a constant ϕ so that $\dot{\phi} = 0$.

(a) Show that the geodesic equation for θ in Prob. 7.3 can be written in the form

$$\frac{d}{dt}(r^2\dot{\theta}) = \frac{2MG}{c^2}\dot{r}\dot{\theta} + O\left(\frac{1}{c^4}\right)$$

(b) Show that the conservation of the angular momentum l in Eq. (8.25) can be recast in the form

$$\frac{d}{dt}(r^2\dot{\theta}) = \frac{2MG}{c^2}\dot{r}\dot{\theta} - \left(\frac{E}{mc^2} + \frac{2MG}{rc^2}\right)\frac{d}{dt}(r^2\dot{\theta}) + O\left(\frac{1}{c^4}\right)$$

where energy conservation in the NRL, that is through $O(c^0)$, has been invoked in the evaluation of the coefficient of the $O(1/c^2)$ term

$$E = \frac{m}{2}(\dot{r}^2 + r^2\dot{\theta}^2) - \frac{mMG}{r} = \text{constant} \qquad ; \text{NRL}$$

(c) Now iterate the result in (b) through $O(1/c^2)$ to reproduce the result in (a).

Solution to Problem 8.4

(a) As in the text, we assume a constant ϕ so that $\dot{\phi} = 0$. The geodesic equation for θ in Prob. 7.3 then take the form

$$\frac{1}{\rho}\frac{d}{dt}(\rho\dot{\theta}) + \frac{2}{r}(\dot{r}\dot{\theta}) = 0$$

In the Schwarzschild metric, through $O(1/c^2)$, the fourth geodesic equation becomes

$$\frac{1}{\rho}\frac{d\rho}{dt} + \frac{2MG}{r^2}\frac{\dot{r}}{c^2} = 0$$

Hence

$$\ddot{\theta} - \frac{2MG}{r^2}\frac{(\dot{r}\dot{\theta})}{c^2} + \frac{2}{r}(\dot{r}\dot{\theta}) = 0$$

Now work out

$$\frac{d}{dt}(r^2\dot{\theta}) = r^2\ddot{\theta} + 2r(\dot{r}\dot{\theta}) = \frac{2MG}{c^2}(\dot{r}\dot{\theta})$$

This establishes the result in part (a)

$$\frac{d}{dt}(r^2\dot\theta) = \frac{2MG}{c^2}\dot r\theta + O\left(\frac{1}{c^4}\right)$$

(b) The conserved angular momentum in Eq. (8.25), derived through lagrangian mechanics, reads

$$l = mr^2\dot\theta\left(1 + \frac{\dot r^2}{2c^2} + \frac{r^2\dot\theta^2}{2c^2} + \frac{MG}{c^2 r}\right) \qquad ; \text{ angular momentum}$$

This result holds through $O(1/c^2)$. Through this order, one can evaluate the coefficient of the $1/c^2$ term using the $O(c^0)$ expression, which is just that in Prob. 8.2

$$T_0 + V_0 = \frac{m}{2}(\dot r^2 + r^2\dot\theta^2) - \frac{mMG}{r} = E = \text{constant}$$

It follows that

$$l = mr^2\dot\theta\left[1 + \frac{1}{mc^2}\left(E + \frac{2mMG}{r}\right)\right] + O\left(\frac{1}{c^4}\right)$$

The statement of conservation of angular momentum then reads

$$\frac{1}{m}\frac{dl}{dt} = \frac{d}{dt}(r^2\dot\theta) - \frac{2MG}{c^2}(\dot\theta\dot r) + \frac{1}{mc^2}\left(E + \frac{2mMG}{r}\right)\frac{d}{dt}(r^2\dot\theta) + O\left(\frac{1}{c^4}\right)$$
$$= 0$$

Therefore

$$\frac{d}{dt}(r^2\dot\theta) = \frac{2MG}{c^2}\dot r\theta - \left(\frac{E}{mc^2} + \frac{2MG}{rc^2}\right)\frac{d}{dt}(r^2\dot\theta) + O\left(\frac{1}{c^4}\right)$$

(c) An iteration through $O(1/c^2)$ of the result in part (b), established through lagrangian mechanics, now reproduce the result in part (a), derived from the geodesic equations

$$\frac{d}{dt}(r^2\dot\theta) = \frac{2MG}{c^2}\dot r\theta + O\left(\frac{1}{c^4}\right)$$

Problem 8.5 Use the following numbers for the earth

$$M_e = 5.98 \times 10^{24}\,\text{kg}$$
$$R_e = 6.38 \times 10^3\,\text{km}$$

Show that the GR contribution to the advance of the perihelion of a satellite in an orbit with very small eccentricity at the surface of the earth is

$$\Omega_{\text{perihelion}} = 16.8''/\text{year} \qquad ; \text{ surface of earth}$$

Solution to Problem 8.5

The formula for the advance of the perihelion is given in Eqs. (8.59)–(8.60)

$$\Delta\theta = 6\pi\lambda = 6\pi \left(\frac{MG}{c^2a}\right) \qquad ; \text{ advance of perihelion}$$

$$\Omega_{\text{perihelion}} = \Delta\theta/\text{rev}$$

This result is correct through $O(\lambda)$ and then exact as the orbit deformation $\eta \to 0$.

The two additional fundamental constants required are then given in Eqs. (8.61)

$$G = 6.673 \times 10^{-20} \text{ km}^3/\text{kg-sec}^2$$

$$c = 2.998 \times 10^5 \text{ km/sec}$$

Thus the advance of the perihelion of a satellite in an orbit with very small eccentricity at the surface of the earth is

$$\Omega_{\text{perihelion}} = 1.31 \times 10^{-8} \text{ rad/rev}$$

Now we just have to adjust the units. The period for circular motion is given in Prob. 7.5

$$p = 2\pi \left(\frac{a^3}{MG}\right)^{1/2}$$

$$= 5.07 \times 10^3 \text{ sec} = 1.61 \times 10^{-4} \text{ year}$$

To convert from radians to degrees

$$2\pi \text{ rad} = 1.30 \times 10^6 \,''$$

Hence

$$\Omega_{\text{perihelion}} = 16.8''/\text{year}$$

which is the stated answer.

Problem 8.6 Consider the *deflection of light in the Schwarzschild metric* (Probs. 8.6–8.7). Work in the plane $\theta = \pi/2$. We are interested in the configuration illustrated in Fig. 8.1, where the light ray moves through the gravitational field in an essentially undeviated straight-line path at an impact parameter b. Define $\Lambda \equiv MG/c^2 b$ and assume $\Lambda \ll 1$.

(a) Start in the LF^3, and show that the interval vanishes for a light signal. Hence, conclude that light follows the null interval $(ds)^2 = 0$ in the global laboratory frame where the Schwarzschild metric applies. Now work in this frame;

(b) Consider first the case where $\Lambda = 0$. Show the vanishing of the interval implies that $v^2/c^2 = 1$ for the light signal (here $v = \dot{x}$);

(c) Consider then the case of small Λ. Show that the vanishing of the interval $(ds)^2 = 0$ implies the existence of an *effective index of refraction* $(v/c)_{\text{eff}} \equiv 1/n$ for light in the gravitational field given by[7]

$$n = 1 + \Lambda \left(\frac{1}{r/b} \right) \left(1 + \frac{\dot{r}^2}{c^2} \right) \qquad ; \text{ effective index of refraction}$$

Thus the deflection of light amounts to tracing a ray through a "medium" with this varying effective index of refraction.

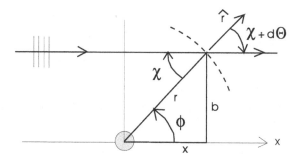

Fig. 8.1 Configuration for deflection of a light ray from an essentially straight-line path at an impact parameter b through the Schwarzschild metric. There is a small change $d\Theta$ in angle of the incident ray relative to the normal $\hat{\mathbf{r}}$ due to the change dn in effective index of refraction. The trajectory extends to infinity in both directions.

---

[7] Here $v^2_{\text{eff}} = \dot{r}^2 + r^2 \dot{\theta}^2 + r^2 \sin^2 \theta \, \dot{\phi}^2$ is the square of the velocity that the record keeper observes on his or her screen.

Solution to Problem 8.6

(a) In the LF^3, one simply has the laws of special relativity, and hence light follows the null interval

$$(ds)^2 = 0 \qquad ; \text{ light in } LF^3$$

Since the physical interval is independent of the coordinate system, light follows this same interval in the global laboratory frame with the Schwarzschild metric and coordinates $q^\mu = (r, \theta, \phi, ct)$

$$(ds)^2 = 0 \qquad ; \text{ light with Schwarzschild metric}$$

(b) Define (see Fig. 8.1)

$$\Lambda \equiv \frac{MG}{c^2 b} \qquad ; \Lambda \ll 1$$

We assume $\Lambda \ll 1$, and work in the (x, y)-plane, with $\theta = \pi/2$. The Schwarzschild metric is then given in Eq. (7.89)

$$(ds)^2 = \frac{(dr)^2}{1 - 2MG/c^2 r} + r^2 (d\phi)^2 - \left(1 - \frac{2MG}{c^2 r}\right)(cdt)^2$$

$$; \text{ Schwarzschild metric, } \theta = \pi/2$$

Start with $\Lambda = 0$. In this case, light follows the interval

$$(ds)^2 = (dr)^2 + r^2 (d\phi)^2 - (cdt)^2 = 0$$

Division by $(dt)^2$ gives

$$\dot{r}^2 + r^2 \dot{\phi}^2 = v^2 = c^2 \qquad ; \Lambda = 0$$

Here we have the familiar identification $v^2 = \dot{r}^2 + r^2 \dot{\phi}^2$.

(c) Now consider the case of small Λ. The vanishing of the interval then gives, through $O(\Lambda)$

$$(ds)^2 = (dr)^2 (1 + 2\Lambda b/r) + r^2 (d\phi)^2 - (1 - 2\Lambda b/r)(cdt)^2 = 0$$

Division by $(cdt)^2$ gives

$$\left(\frac{v}{c}\right)^2_{\text{eff}} - 1 + \frac{2\Lambda b}{r}\left(1 + \frac{\dot{r}^2}{c^2}\right) = 0$$

Here we have defined

$$(v)^2_{\text{eff}} \equiv \dot{r}^2 + r^2 \dot{\phi}^2$$

This is now the square of the *effective velocity of light* that the record keeper sees on his or her screen with the Schwarzschild metric.

Define an *index of refraction n* by

$$\left(\frac{v}{c}\right)_{\text{eff}} \equiv \frac{1}{n} \qquad ; \text{ effective index of refraction}$$

Therefore, through $O(\Lambda)$,

$$n = 1 + \Lambda \left(\frac{1}{r/b}\right)\left(1 + \frac{\dot{r}^2}{c^2}\right)$$

Hence the deflection of light amounts to tracing a ray through a "medium" with this varying effective index of refraction.

Problem 8.7 Since $r \approx x$ for most of the trajectory in Fig. 8.1, with $\dot{r}^2/c^2 \approx \dot{x}^2/c^2 \approx 1$, assume the contribution of the second term in the final parentheses in Prob. 8.6(c) is approximately equal to that of the first. More generally, write its contribution as γ times that of the first (with $\gamma \approx 1$). The effective index of refraction then takes the approximate form

$$n \approx 1 + 2\Lambda \left(\frac{1}{r/b}\right)\left(\frac{1+\gamma}{2}\right) \qquad ; \text{ effective index of refraction}$$

The advantage of making this approximation is that one now knows that the surfaces of constant n are spheres with the normal to the surface given by $\hat{\mathbf{r}}$. The problem is then reduced to an application of *Snell's law* from freshman physics. Snell's law states that at the interface between two media with different indices of refraction, a light ray is refracted according to the relation $n_i \sin \chi_i = n_r \sin \chi_r$, where the angles are measured relative to the normal to the surface.

(a) Show the differential form of Snell's law as $n \to n + dn$ and $\chi \to \chi + d\Theta$ is (see Fig. 8.1)[8]

$$d\Theta = -\frac{dn}{n} \tan \chi$$

(b) With the observation that $r^2 = x^2 + b^2$, and the definition $u \equiv x^2/b^2$, show that the relation in (a) becomes

$$d\Theta = -\left[\frac{1}{n}\frac{dn}{du} \tan \chi\right] du \qquad ; u \equiv \frac{x^2}{b^2}$$

$$= \Lambda \frac{(1+\gamma)}{2}\left[\frac{1}{(1+u)^{3/2}} \tan \chi\right] du + O(\Lambda^2)$$

---

[8]When entering a medium with higher effective index of refraction, the ray is bent *toward* the normal; when entering one with lower effective index of refraction, it is bent *away* from the normal.

(c) Show from Fig. 8.1 that

$$\tan \chi = \tan \phi = \frac{b}{x} = \frac{1}{\sqrt{u}}$$

(d) Now integrate the relation in (b) from 0 to ∞ to get the net change Θ_0 from the outgoing leg of the trajectory

$$\Theta_0 = 2\Lambda \left(\frac{1+\gamma}{2} \right)$$

(e) Show there is an equivalent deflection from the incoming leg, and hence obtain the total deflection of the light ray through $O(\Lambda)$

$$\Theta_{\text{deflect}} = 2\Theta_0 = 4\Lambda \left(\frac{1+\gamma}{2} \right) \qquad ; \Lambda \equiv \frac{MG}{c^2 b}$$

This is the result given in [Weinberg (1972)] for the deflection of light in the Schwarzchild metric through $O(\Lambda)$, where it is stated that $\gamma = 1$ is actually an exact result.[9]

Solution to Problem 8.7

(a) We refer to Fig. 8.1. On the outgoing leg, as the ray moves to region of lower index of refraction, it is bent away from the normal (positive $d\Theta$). Snell's law says

$$n_f \sin \chi_f = n_i \sin \chi_i$$

The differential form of this relation is

$$(n + dn) \sin (\chi + d\Theta) \approx (n + dn)(\sin \chi + d\Theta \, \cos \chi) = n \sin \chi$$

$$d\Theta = -\frac{dn}{n} \tan \chi$$

(b) Observe from Fig. 8.1 that $r^2 = x^2 + b^2$, where x is the x-coordinate and b is the impact parameter. Introduce the quantity u

$$u \equiv \frac{x^2}{b^2} \qquad ; \left(\frac{r}{b} \right)^2 = 1 + u$$

and use

$$n = 1 + 2\Lambda \left(\frac{1}{\sqrt{1+u}} \right) \left(\frac{1+\gamma}{2} \right)$$

---

[9] For some accurate analytical formulas for gravitational lensing that follow from the result derived in [Weinberg (1972)], see [Amore and Arceo (2006)].

The relation in part (a) then becomes

$$d\Theta = -\left[\frac{1}{n}\frac{dn}{du}\tan\chi\right]du$$

$$= \Lambda\frac{(1+\gamma)}{2}\left[\frac{1}{(1+u)^{3/2}}\tan\chi\right]du + O(\Lambda^2)$$

(c) It follows from Fig. 8.1 that

$$\tan\chi = \tan\phi = \frac{b}{x} = \frac{1}{\sqrt{u}}$$

(d) To get the net change Θ_0 from the outgoing leg of the trajectory, one can now simply integrate the relation in (b) on u from 0 to ∞, using[10]

$$\int_0^\infty \frac{du}{\sqrt{u}}\frac{1}{(1+u)^{3/2}} = 2$$

$$\Theta_0 = 2\Lambda\left(\frac{1+\gamma}{2}\right)$$

(e) On the incoming leg, the wave is bent toward the normal $-\hat{\mathbf{r}}$ through an angle

$$d\Theta = \left[\frac{1}{n}\frac{dn}{du}\tan\bar{\chi}\right]du$$

where

$$\phi + \bar{\chi} = \pi \qquad ; \tan\bar{\chi} = \frac{b}{|x|} = \frac{1}{\sqrt{u}}$$

Then with

$$-\int_\infty^0 \frac{du}{\sqrt{u}}\frac{1}{(1+u)^{3/2}} = 2$$

one obtains exactly the same deflection as on the outgoing leg. Hence, the total deflection of the light ray through $O(\Lambda)$ is

$$\Theta_{\text{deflect}} = 2\Theta_0 = 4\Lambda\left(\frac{1+\gamma}{2}\right) \qquad ; \Lambda \equiv \frac{MG}{c^2 b}$$

---

[10]Introduce $u = \tan^2\theta$, then

$$\int_0^\infty \frac{du}{\sqrt{u}}\frac{1}{(1+u)^{3/2}} = 2\int_0^{\pi/2}\frac{\tan\theta\sec^2\theta\,d\theta}{\tan\theta}\frac{1}{\sec^3\theta} = 2\int_0^{\pi/2}\cos\theta\,d\theta = 2$$

This is the result given in [Weinberg (1972)] for the deflection of light in the Schwarzchild metric through $O(\Lambda)$, where it is stated that $\gamma = 1$ is actually an exact result.

Note that the total deflection angle is positive, indicating an attractive gravitational interaction between the light and the massive body by which it is passing.

Problem 8.8 This problem reviews some results for orbital motion in the newtonian limit. Start from $L = L_{\text{NRL}}$ in Eq. (8.15) and use the spherical unit vectors defined in Fig. 5.4 in the text.

(a) Show the canonical angular momenta are given by

$$p_\phi = \frac{\partial L}{\partial \dot\phi} = mr^2\dot\phi\,\sin^2\theta \qquad ; \ \dot p_\phi = 0$$

$$p_\theta = \frac{\partial L}{\partial \dot\theta} = mr^2\dot\theta \qquad\qquad ; \ \dot p_\theta = p_\phi\dot\phi\cot\theta$$

(b) Define the angular momentum in this limit by $\vec{l} = \vec{r} \times (m\vec{v})$ where $\vec{v} = \dot r\,\hat{e}_r + r\dot\theta\,\hat{e}_\theta + r\dot\phi\sin\theta\,\hat{e}_\phi$. Show

$$\hat{e}_\phi \cdot \vec{l} = p_\theta \qquad\qquad ; \ \hat{e}_z \cdot \vec{l} = p_\phi$$

where $\hat{e}_z$ is a unit vector in the z-direction. Hence verify the claims made in Figs. 8.3 and 8.4 in the text.

Solution to Problem 8.8

(a) The lagrangian $L_{\text{NRL}}(r, \theta, \phi; \dot r, \dot\theta, \dot\phi)$ in Eqs. (8.15) and (8.12) is

$$L_{\text{NRL}} = \frac{m}{2}\left(\dot r^2 + r^2\dot\theta^2 + r^2\sin^2\theta\,\dot\phi^2\right) + \frac{mMG}{r}$$

Lagrange's equations are

$$\frac{d}{dt}\frac{\partial L}{\partial(\partial L/\partial \dot q^i)} - \frac{\partial L}{\partial q^i} = 0 \qquad ; \ q^i = (r, \theta, \phi)$$

The canonical momenta are

$$p_i = \frac{\partial L}{\partial(\partial L/\partial \dot q^i)}$$

Since ϕ is cyclic (it does not appear in L) one has

$$p_\phi = mr^2\dot\phi\sin^2\theta$$
$$\dot p_\phi = 0$$

For θ, one obtains

$$p_\theta = mr^2\dot\theta$$
$$\dot p_\theta = mr^2\dot\phi^2\sin\theta\cos\theta = p_\phi\dot\phi\cot\theta$$

(b) Define the angular momentum in this limit by

$$\vec l = \vec r\times(m\vec v)$$
$$\vec r = r\,\hat{\mathbf e}_r$$
$$\vec v = \dot r\,\hat{\mathbf e}_r + r\dot\theta\,\hat{\mathbf e}_\theta + r\dot\phi\sin\theta\,\hat{\mathbf e}_\phi$$

where the orthonormal spherical unit vectors are shown in Fig. 5.4 in the text. Then

$$\vec l = mr^2\dot\theta\,\hat{\mathbf e}_\phi - mr^2\dot\phi\sin\theta\,\hat{\mathbf e}_\theta$$

Hence

$$\hat{\mathbf e}_\phi\cdot\vec l = mr^2\dot\theta = p_\theta$$

Let $\hat{\mathbf e}_z$ be a unit vector in the z-direction. Then from Fig. 5.4 in the text

$$\hat{\mathbf e}_z\cdot\hat{\mathbf e}_\phi = 0 \qquad ;\ \hat{\mathbf e}_z\cdot\hat{\mathbf e}_\theta = -\sin\theta$$

It follows that

$$\hat{\mathbf e}_z\cdot\vec l = mr^2\dot\phi\sin^2\theta = p_\phi$$

These results serve as justification for the statements made below Eqs. (8.18) and (8.20): "Recall the newtonian limit (see Prob. 8.8). In that limit, p_ϕ is the component of the angular momentum along the z-axis. One can choose to put the total angular momentum vector in the (x, y) plane, and then $p_\phi = 0$, as illustrated in Fig. 8.3 in the text.[11] The motion then takes place in a plane perpendicular to the angular momentum. We redraw the resulting planar configuration in Fig. 8.4 in the text."

_____

[11]In this case $\dot\phi = 0$, and $\phi = \phi_0 =$ constant [compare Eqs. (8.20)]; this implies $\dot p_\theta = 0$, and $p_\theta =$ constant.

Problem 8.9 (a) Expand the lagrangian in Eq. (8.9) as a power series in $2\Phi/c^2$ for *any* $v^2 = \dot{r}^2 + r^2\dot{\theta}^2 + r^2\sin^2\theta\,\dot{\phi}^2$ with $v^2/c^2 < 1$. Show that through first order, L takes the form

$$L = -mc^2(1-\beta^2)^{1/2} - m\Phi(r)\frac{1+\dot{r}^2/c^2}{(1-\beta^2)^{1/2}} \qquad ; \vec{\beta} = \vec{v}/c$$

This provides one consistent extension of the free relativistic lagrangian in Eq. (6.36), which includes an interaction.

(b) To simplify things, neglect the $\dot{r}^2/c^2$ in the numerator of the second term (this does preserve the NRL). Then write $\vec{\beta} = d\vec{x}/d(ct)$ and use cartesian coordinates. What is the contribution of the interaction to the relativistic form of Newton's second law in the first of Eqs. (6.30)?

(c) Compare with the result in Prob. 6.2.

Solution to Problem 8.9

(a) The lagrangian for particle motion in the Schwarzschild metric in Eq. (8.9) is

$$L = -mc^2\left\{\left[1 + \frac{2\Phi(r)}{c^2}\right] - \frac{1}{c^2}\left[\frac{\dot{r}^2}{1+2\Phi(r)/c^2} + r^2\dot{\theta}^2 + r^2\sin^2\theta\,\dot{\phi}^2\right]\right\}^{1/2}$$

where $\Phi(r) = -MG/r$ is the gravitational potential. Expand this lagrangian as a power series in $2\Phi/c^2$. First, expand inside the square root

$$L \approx -mc^2\left[1 - \frac{v^2}{c^2} + \frac{2\Phi(r)}{c^2}\left(1 + \frac{\dot{r}^2}{c^2}\right)\right]^{1/2}$$
$$v^2 = \dot{r}^2 + r^2\dot{\theta}^2 + r^2\sin^2\theta\,\dot{\phi}^2$$

Then, through first order in $2\Phi/c^2$, the lagrangian becomes,

$$L \approx -mc^2(1-\beta^2)^{1/2} - m\Phi(r)\frac{1+\dot{r}^2/c^2}{(1-\beta^2)^{1/2}} \qquad ; \vec{\beta} = \frac{\vec{v}}{c}$$

(b) Suppose, as instructed, we neglect the $\dot{r}^2/c^2$ term.[12] Then with the introduction of cartesian coordinates, the lagrangian $L(x,y,z;\dot{x},\dot{y},\dot{z})$

---

[12]This is clearly valid in the NRL. In the extreme relativistic limit (ERL) where $\dot{r}^2 \approx c^2$, one could replace $1+\dot{r}^2/c^2 \approx 2$ (see Prob. 8.7).

is given by

$$L \approx -mc^2(1 - \beta^2)^{1/2} - \frac{m\Phi(r)}{(1 - \beta^2)^{1/2}}$$

$$\beta^2 = \frac{1}{c^2}(\dot{x}^2 + \dot{y}^2 + \dot{z}^2) \qquad ; r = (x^2 + y^2 + z^2)^{1/2}$$

The canonical momentum p_x is now obtained as[13]

$$p_x = \frac{\partial L}{\partial \dot{x}} = \frac{mv_x}{(1 - \beta^2)^{1/2}} - \frac{m\Phi}{c^2} \frac{v_x}{(1 - \beta^2)^{3/2}}$$

Hence

$$\vec{p} = \frac{m\vec{v}}{(1 - \beta^2)^{1/2}} \left[1 - \frac{1}{(1 - \beta^2)} \frac{\Phi}{c^2} \right]$$

Lagrange's equations then give

$$\frac{d\vec{p}}{dt} = -\frac{m}{(1 - \beta^2)^{1/2}} \boldsymbol{\nabla}\Phi$$

Newton's second law for a free, relativistic particle in the first of Eqs. (6.30) reads

$$\frac{d\vec{p}}{dt} = \frac{d}{dt} \left[\frac{m\vec{v}}{(1 - \beta^2)^{1/2}} \right] = 0 \qquad ; \text{Newton's law}$$

We see the following modifications with our new equation of motion in the gravitational field:

- The gravitational field now exerts a force $-\boldsymbol{\nabla}\Phi$;
- It is the relativistic mass $m/(1-\beta^2)^{1/2}$ that enters into the coupling to the gravitational field;
- Due to the velocity-dependent coupling to the field in L, the kinetic momentum appearing in Newton's law is modified by a factor $[1 - \Phi/(1 - \beta^2)c^2]$.

(c) In Prob. 6.2 the relativistic particle lagrangian was simply augmented with a static potential $V(\vec{x})$. Newton's second law then reads

$$\frac{d\vec{p}}{dt} = \frac{d}{dt} \left[\frac{m\vec{v}}{(1 - \beta^2)^{1/2}} \right] = -\vec{\nabla}V$$

---

[13] Use $\partial \beta / \partial \dot{x} = v_x / \beta c^2$.

Problem 8.10 Suppose the lagrangian L for the motion of a point particle of mass m in the Schwarzschild metric in Eq. (8.10) is augmented with a potential $V(r)$ as in Prob. 6.2. Expand through $O(1/c^2)$. How are the subsequent equations of motion in the text modified?

Solution to Problem 8.10

The lagrangian for particle motion in the Schwarzschild metric is given in Eq. (8.9). The substitution $L \to L - V(r)$ simply adds $-V(r)$ to the r.h.s. of that expression. The expansion of the lagrangian through $O(1/c^2)$ is then carried out in the text. Since the additional term is independent of angles, the angular analysis is unchanged, and the resulting lagrangian is thus given in Eq. (8.22), with an additional $-V(r)$ on the r.h.s. The angular analysis in Eqs. (8.23)–(8.31) following from that lagrangian remains unchanged; however, the radial Eq. (8.32) now has an additional force $-dV(r)/dr$ on the r.h.s.

Chapter 9

Gravitational Redshift

Problem 9.1 (a) Show through $O(1/c^2)$ that the shift in frequency of a photon traveling up a small distance h at the earth's surface can be written as

$$\frac{\Delta\nu}{\nu} = -\frac{gh}{c^2}$$

where g is the acceleration due to gravity.

(b) Take $g = 9.807\,\text{m/sec}^2$ and assume $h = 10\,\text{m}$. Compute the fractional shift in frequency of a spectral line.

Solution to Problem 9.1

(a) The fractional shift in photon frequency is calculated in Eqs. (9.17) and (9.11)

$$\frac{\Delta\nu}{\nu} = -\frac{\Delta\Phi}{c^2}$$

$$\Delta\Phi = -MG\left(\frac{1}{r_2} - \frac{1}{r_1}\right) \approx MG\frac{\Delta r}{r^2} \qquad ; \Delta r \equiv r_2 - r_1$$

The last relation holds provided $|\Delta r| \ll r$.

Now identify the height above the earth's surface $h = r - R_e = \Delta r$, where R_e is the earth's radius, and the force/(unit mass) at the earth's surface as $g = M_e G/R_e^2$ where M_e is the earth's mass[1]

$$h = r - R_e \qquad ; g = \frac{M_e G}{R_e^2}$$

---

[1] See Prob. 8.5.

Then, through $O(1/c^2)$

$$\frac{\Delta\nu}{\nu} \approx -\frac{gh}{c^2}$$

(b) With $g = 9.807\,\text{m/sec}^2$, $c = 2.998 \times 10^8\,\text{m/sec}$, and $h = 10\,\text{m}$, one obtains

$$\frac{\Delta\nu}{\nu} = -1.09 \times 10^{-15}$$

The frequency gets lower as one moves up.

Problem 9.2 The lifetime of a free muon is $\tau = 2.197 \times 10^{-6}\,\text{sec}$. How close to the Schwarzschild radius, $r/R_s = 1 + \epsilon$, must a muon be to have its laboratory lifetime extended to 1 sec?

Solution to Problem 9.2

The time dilation in the Schwarzschild metric is calculated in Eq. (7.109). If τ is the lifetime in the LF^3, and τ_L the lifetime in the lab frame, then

$$\tau_L = \frac{\tau}{(1 - R_s/r)^{1/2}}$$

This is re-written as

$$\frac{r}{R_s} = \frac{1}{1 - (\tau/\tau_L)^2}$$

With $r/R_s = 1 + \epsilon$, this becomes

$$\epsilon = \frac{(\tau/\tau_L)^2}{1 - (\tau/\tau_L)^2}$$

Now take the lifetime of a free muon $\tau = 2.197 \times 10^{-6}\,\text{sec}$, and extend it to $\tau_L = 1\,\text{sec}$. Then

$$\epsilon \approx 4.83 \times 10^{-12}$$

This is how close to the Schwarzschild radius $r/R_s = 1 + \epsilon$ the muon must get to have the record keeper observe a lifetime of 1 second.

Problem 9.3 A few spectral lines thought to come from the surface of a distant body are observed to have a gravitational redshift of $\Delta\nu/\nu \approx -10^{-6}$. Calculate the ratio of mass to radius for that body.

Solution to Problem 9.3

From Eq. (9.17), the fractional shift in frequency is related to the difference in gravitational potential by

$$\frac{\Delta \nu}{\nu} = -\frac{\Delta \Phi}{c^2}$$
$$= \frac{MG}{c^2}\left(\frac{1}{r_2} - \frac{1}{r_1}\right)$$

Now take $r_1 = R$, and $r_2 = \infty$, for light emitted from the surface and
observed far away, to arrive at

$$\frac{\Delta \nu}{\nu} = -\frac{G}{c^2}\left(\frac{M}{R}\right) \qquad ; (r_1, r_2) = (R, \infty)$$

From Prob. 8.5

$$G = 6.673 \times 10^{-20} \text{ km}^3/\text{kg-sec}^2$$
$$c = 2.998 \times 10^5 \text{ km/sec}$$

Thus, with $\Delta \nu / \nu \approx -10^{-6}$, the ratio of mass to radius for the body is

$$\frac{M}{R} = -\frac{c^2}{G}\frac{\Delta \nu}{\nu} = 1.35 \times 10^{24}\,\frac{\text{kg}}{\text{km}} \qquad ; \frac{\Delta \nu}{\nu} \approx -10^{-6}$$

Problem 9.4 Problems 9.4–9.8 examine the radial propagation of light in
the Schwarzschild metric.[2] Consider a *classical electromagnetic wave*. We
first make some observations concerning the wavelength of this wave in the
gravitational potential.

(a) From the fact that light follows the null interval $(ds)^2 = 0$ in the LF^3
[frame III], conclude that for propagation in the radial direction, the light
signal must satisfy the following differential relation in the global inertial
laboratory frame I

$$dr = \left(1 + \frac{2\Phi}{c^2}\right)c\,dt$$

(b) Define dN as a given number of oscillations of the electric field at
the origin in the LF^3 at a given point. Further define the proper time
element $d\tau$ and proper frequency $\bar{\nu} = dN/d\tau$ in that frame. Let $d\bar{l}$ be an
element of proper radial distance traversed by the light signal in the time
$d\tau$ in the LF^3. Substitute the relations in Eqs. (7.109) and (7.113) between

---

[2]In these problems, λ represents the wavelength.

laboratory quantities and proper quantities into the expression in (a), and verify the following

$$d\bar{l} = c\,d\tau$$

$$\bar{\lambda} = \frac{d\bar{l}}{dN} = \frac{c}{\bar{\nu}}$$

Hence, one indeed recovers the correct relation between wavelength and frequency in the LF^3 from (a).

(c) In direct analogy to the analysis in (b), define the wavelength for a light signal propagating in the radial direction in the Schwarzschild metric in the global inertial laboratory frame I by $\lambda \equiv dr/dN$. Show from (a) that

$$\lambda = \frac{dr}{dN} = \bar{\lambda}\left(1 + \frac{2\Phi}{c^2}\right)^{1/2}$$

Solution to Problem 9.4

(a) Light follows the null interval in the LF^3. It then follows this same physical interval in the global laboratory frame

$$(d\mathbf{s})^2 = 0 \qquad\qquad ; \text{ light}$$

With only radial motion, the interval in the Schwarzschild metric in Eq. (7.89) gives

$$(d\mathbf{s})^2 = \frac{(dr)^2}{1 - (2MG/c^2 r)} - \left(1 - \frac{2MG}{c^2 r}\right)(cdt)^2 = 0$$

Hence

$$dr = \left(1 + \frac{2\Phi}{c^2}\right)c\,dt \qquad\qquad ; \Phi = -\frac{MG}{r}$$

(b) Equations (7.109) and (7.113) relate the laboratory time and radial distance increments (dt, dl) to the proper quantities $(d\tau, d\bar{l})$ [3]

$$dt = \frac{d\tau}{(1 + 2\Phi/c^2)^{1/2}}$$

$$dl = \left(1 + 2\Phi/c^2\right)^{1/2} d\bar{l}$$

Note both the time dilation and length contraction. It then follows from the relation in part (a) that one then recovers the correct expression for the

---

[3] Note $R_s/r = -2\Phi/c^2$; now $dl \equiv dr$ is an element of radial distance.

distance travelled by light in the LF[3]

$$\bar{dl} = c\, d\tau$$

If dN is a given number of oscillations of the electric field at the origin in the LF[3], at a given point, and the proper frequency is $\bar{\nu} = dN/d\tau$, then the proper wavelength is given by

$$\bar{\lambda} = \frac{\bar{dl}}{dN} = \frac{\bar{dl}}{d\tau}\frac{d\tau}{dN} = \frac{c}{\bar{\nu}}$$

One thus also recovers the correct expression for the wavelength of the light in the LF[3].

(c) The wavelength of the light in the laboratory frame is given by

$$\lambda = \frac{dl}{dN} = \frac{dl}{\bar{dl}}\frac{\bar{dl}}{dN} = \bar{\lambda}\left(1 + \frac{2\Phi}{c^2}\right)^{1/2}$$

Problem 9.5 In exactly the same manner as in Prob. 9.4(c), derive the following relations for the frequency defined by $\nu \equiv dN/dt$

$$\nu\lambda = c\left(1 + \frac{2\Phi}{c^2}\right)$$

$$\nu = \bar{\nu}\left(1 + \frac{2\Phi}{c^2}\right)^{1/2}$$

Now note that the fractional change in this frequency when the gravitational potential is changed by $\Delta\Phi$ is given through $O(1/c^2)$ by

$$\frac{\Delta\nu}{\nu} = \frac{\Delta\Phi}{c^2} \qquad ;\ \text{oscillator } (1) \to (2)$$

Note the sign. This expression is identical to that derived in the text for the frequency of an oscillator in the gravitational field. The difficulty in attempting to apply the analysis in this problem (and in the preceding one) to the *propagation* of a light signal in the Schwarzschild metric is that it *does not fully take into account the effect of the gravitational potential on the frequency of the photon*. This is addressed in the text, and the correct answer for the frequency shift through $O(1/c^2)$ is that given in Eq. (9.21).

Solution to Problem 9.5

As in Prob. 9.4(c), the frequency of the light in the laboratory frame is given by

$$\nu = \frac{dN}{dt} = \frac{dN}{d\tau}\frac{d\tau}{dt} = \bar{\nu}\left(1 + \frac{2\Phi}{c^2}\right)^{1/2}$$

This gives[4]

$$\nu\lambda = \bar{\nu}\bar{\lambda}\left(1 + \frac{2\Phi}{c^2}\right) = c\left(1 + \frac{2\Phi}{c^2}\right)$$

Readers are here referred back to the discussion in the statement of the problem.

Problem 9.6 Let us examine what the *record keeper sees* with a radially propagating light signal in the Schwarzschild metric.

(a) Start in the LF^3 where the velocity of light is c, and demonstrate that the light signal follows the null interval $(ds)^2 = 0$ in the global inertial laboratory frame I in the Schwarzschild metric.

(b) The record keeper logs the coordinates (r, ct) for the light signal. Show that what he or she observes for all $r \geq R_s$ is that

$$\frac{dr}{d(ct)} = \left(1 + \frac{2\Phi}{c^2}\right)$$

(c) Note that $\dot{r} \leq c$. What does the record keeper observe for $\dot{r}$ at the Schwarzschild radius?

Solution to Problem 9.6

(a) As in Prob. 9.4(a), the physical interval for light in the LF^3 is the "null interval", satisfying $(ds)^2 = 0$. This is then also the physical interval in the laboratory frame.

(b) The record keeper logs the coordinates $q^\mu = (r, \theta, \phi, ct)$. For radial motion, light thus satisfies the relation derived in Prob. 9.4(a)

$$\frac{1}{c}\frac{dr}{dt} = \left(1 + \frac{2\Phi}{c^2}\right) = \left(1 - \frac{R_s}{r}\right)$$

This is the *effective speed of light* that the record keeper observes for radial propagation. Note that $\dot{r} < c$.

(c) At the Schwarzschild radius, this *vanishes!*

---

[4]Note this is *no longer* $\nu\lambda = c$.

Problem 9.7 Problems 9.4 and 9.5 *do* provide the wavelength and frequency for a classical electromagnetic wave at a given point in the global inertial laboratory frame I in the Schwarzschild metric in terms of the corresponding quantities in the LF^3.

(a) Show

$$\nu\lambda = \dot{r}$$

where $\dot{r}$ is the radial velocity observed by the record keeper in Prob. 9.6. Discuss.

(b) Show that when combined with the correct expression in Eq. (9.21) for $\Delta\nu/\nu$ for a propagating light signal, this gives the corresponding wavelength shift as

$$\frac{\Delta\lambda}{\lambda} = 3\,\frac{\Delta\Phi}{c^2} \qquad ; \text{ wavelength } (1) \to (2)$$

Again, to this order, one can use either λ_1 or λ_2 in the denominator on the l.h.s.[5] As the light wave moves to a region of higher potential $\Delta\Phi > 0$, the wavelength increases $\Delta\lambda > 0$. This is the *gravitational redshift*.

Solution to Problem 9.7

(a) From Probs. 9.4–9.5, for propagation of light in the radial direction in the Schwarzschild metric

$$\lambda = \bar{\lambda}\left(1 + \frac{2\Phi}{c^2}\right)^{1/2}$$

$$\nu = \bar{\nu}\left(1 + \frac{2\Phi}{c^2}\right)^{1/2}$$

Hence

$$\nu\lambda = c\left(1 + \frac{2\Phi}{c^2}\right) = \dot{r}$$

where we have used $\bar{\nu}\bar{\lambda} = c$, and the result for $\dot{r}$ in Prob. 9.6(b). Thus the record keeper sees precisely the correct relation between frequency and wavelength for a wave of velocity $\dot{r}$.

(b) To first order in small differences, the above relation becomes

$$\lambda\,\Delta\nu + \nu\,\Delta\lambda = \frac{2\Delta\Phi}{c}$$

---

[5] Or even $\bar{\lambda}$.

Division by $\nu\lambda$ allows this to be re-written, through $O(1/c^2)$, as

$$\frac{\Delta\nu}{\nu} + \frac{\Delta\lambda}{\lambda} \approx 2\frac{\Delta\Phi}{c^2}$$

The correct expression for $\Delta\nu/\nu$ for a propagating light signal in Eq. (9.21) is

$$\frac{\Delta\nu}{\nu} = -\frac{\Delta\Phi}{c^2}$$

It follows that

$$\frac{\Delta\lambda}{\lambda} \approx 3\frac{\Delta\Phi}{c^2}$$

which is the stated answer for the *gravitational redshift*.

Problem 9.8 Suppose there were to be an (uncalculated) energy shift $\Delta\varepsilon/\varepsilon = O(1/c^2)$ of the atom or nucleus between points (1) and (2), or between the lab and LF^3, in Fig. 9.3 in the text. Show the results in Eqs. (9.20) and (9.21) still hold through the stated order.

Solution to Problem 9.8

Suppose one modifies Eq. (9.18) to read

$$E_2 = m_A c^2 + \varepsilon + \Delta\varepsilon \equiv m_A^{\star}c^2$$

where $\Delta\varepsilon$ is some additional, uncalculated energy shift of order $1/c^2$. Then Eqs. (9.19)–(9.21), which are proportional to ε/c^2, continue to hold through the stated order of $1/c^2$. [See the comment after Eq. (9.21).]

Chapter 10

Neutron Stars

Problem 10.1 Work in the newtonian limit, and assume an equation of state $P(\rho)$.

(a) Use Gauss's law to obtained the gravitational field as a function of r inside a spherically symmetric mass distribution;

(b) Balance the pressure force and gravitational attraction on a small mass element to obtain the newtonian expression for stellar structure in Eq. (10.60).

Solution to Problem 10.1

(a) Just as in electrostatics, the $1/r^2$ gravitational field inside a spherically symmetric mass distribution is obtained from the mass inside a sphere of radius r, acting as a point particle at the origin. Thus

$$\vec{g}(r) = -\frac{G\mathcal{M}(r)}{r^2}\frac{\vec{r}}{r}$$

$$\mathcal{M}(r) = \int_0^r 4\pi s^2 \rho(s)\, ds$$

The gravitational force on a small mass element ρdV at a radius r is then given by

$$\vec{F}_{\text{grav}} = \vec{g}(r)\, \rho dV$$

(b) The pressure force on a volume dV is given by

$$\vec{F}_{\text{press}} = -\vec{\nabla} P\, dV$$

If everything is radial, then

$$\vec{F}_{\text{press}} = -\frac{dP(r)}{dr}\frac{\vec{r}}{r}\, dV$$

In equilibrium, the gravitational force and pressure force must balance

$$\vec{F}_{\text{grav}} + \vec{F}_{\text{press}} = 0 \qquad ; \text{equilibrium}$$

Hence

$$\frac{dP}{dr} = -\frac{G\mathcal{M}(r)\rho(r)}{r^2} \qquad ; \text{newtonian limit}$$

$$\mathcal{M}(r) = \int_0^r 4\pi s^2 \rho(s)\,ds$$

These are the equations for stellar structure in the newtonian limit. They comprise the $c^2 \to \infty$ limit of the TOV Eqs. (10.76)–(10.77). It remains to augment them with the equation of state $P(\rho)$.

Problem 10.2 Given the fact that the mass density $\rho(r)$ is a positive quantity which decreases with r, and that $\rho(0)$ is finite, show that the inequality in Eq. (10.71) implies that the denominator in the integral in Eq. (10.68) will not vanish.

Solution to Problem 10.2

The denominator in Eq. (10.68) is

$$\mathcal{D}(r) = 1 - \frac{2G\mathcal{M}(r)}{c^2 r}$$

The mass $\mathcal{M}(r)$, star radius R, and star mass M are given by Eqs. (10.60)–(10.63),

$$\mathcal{M}(r) = \int_0^r 4\pi s^2 \rho(s)\,ds \qquad ; \rho(R) = 0$$

$$\mathcal{M}(R) = M$$

Consider the quantity

$$r\mathcal{D}(r) = r - \frac{R_s}{M}\int_0^r 4\pi s^2 \rho(s)\,ds \qquad ; R_s = \frac{2MG}{c^2}$$

where R_s is the Schwarzschild radius and r lies in the interval $[0, R]$. We are interested here in the case where the star radius R is larger than the Schwarzschild radius [see Eq. (10.71)]

$$R > R_s$$

For $r \to 0$, one has

$$rD(r) \to r - \frac{R_s}{M} \frac{4\pi r^3}{3} \rho(0) \qquad ; r \to 0$$

For $r \to R$, one finds

$$rD(r) \to R - R_s \qquad ; r \to R$$

Thus, both limits of $rD(r)$ on the interval $[0, R]$ are positive.
   For orientation, take a model uniform distribution

$$\rho(r) = \rho(0) \qquad ; 0 \le r \le R$$

This gives

$$rD(r) = r - R_s \left(\frac{r}{R}\right)^3$$

The condition that the denominator not vanish in the interval, $rD(r) > 0$,
then implies

$$\frac{R}{R_s} > \left(\frac{r}{R}\right)^2 \qquad ; rD(r) > 0$$

Given $R > R_s$, this holds for all r in the interval $[0, R]$.
   To model a smooth fall-off in the surface, take a mass density

$$\rho(r) = \rho(0) \left[1 - \left(\frac{r}{R}\right)^n\right] \qquad ; 0 \le r \le R$$

This gives

$$\mathcal{M}(r) = \frac{4\pi r^3 \rho(0)}{3} \left[1 - \frac{3}{n+3} \left(\frac{r}{R}\right)^n\right]$$

and

$$rD(r) = r - \frac{R_s}{[1 - 3/(n+3)]} \left(\frac{r}{R}\right)^3 \left[1 - \frac{3}{n+3} \left(\frac{r}{R}\right)^n\right]$$

In this case, the condition $rD(r) > 0$ implies

$$\frac{R}{R_s} \left[1 - \frac{3}{n+3}\right] > \left(\frac{r}{R}\right)^2 \left[1 - \frac{3}{n+3} \left(\frac{r}{R}\right)^n\right]$$

A calculation with Mathcad 7 indicates that for $n = 8$, this holds everywhere on the interval $[0, R]$ for R down to within 1.8% of R_s.
   In Table 10.1 we show the ratio $(R/R_s)_{\min}$ for which $rD(r) > 0$ in the interval $[0, R]$ for a few n. The claim in the problem is actually invalid, as

this artificial $\rho(r)$ provides a counter example; however, one still gets down very close to $R = R_s$.

Table 10.1   The ratio $(R/R_s)_{min}$ for a few n.

n	$(R/R_s)_{min}$
8	1.018
10	1.015
12	1.013

The situation as $R \to R_s$ with a more realistic $\rho(r)$ generated as a solution to the TOV equations is discussed in detail in the text.

Problem 10.3 (a) Write, or obtain, a program to solve the TOV Eqs. (10.76) numerically.

(b) Choose the "stiff" equation of state $P = \rho c^2$. Find (R, M) for a few neutron stars.[1]

(c) What is the maximum mass of a neutron star in units of $M_\odot$ with this equation of state?

Solution to Problem 10.3

(a) The TOV Eqs. (10.76) for the neutron star are

$$\frac{dP}{dr} = -\frac{G\mathcal{M}(r)\rho(r)}{r^2}\left[\frac{1 + P(r)/\rho(r)c^2}{1 - 2G\mathcal{M}(r)/c^2 r}\right]\left[1 + \frac{4\pi P(r)r^3}{\mathcal{M}(r)c^2}\right]$$

$$\mathcal{M}(r) = \int_0^r 4\pi s^2 \rho(s)\, ds$$

$$P = P(\rho) \qquad\qquad ; \text{ equation of state}$$

The supplied equation of state (EOS) expresses the pressure in terms of the mass density at each point. The boundary conditions that then go with this non-linear, integro-differential equation for the radial dependence of the pressure $P(r)$ are those in Eqs. (10.61) and (10.81)

$$P(R) = 0 \qquad\qquad ; \text{ defines surface}$$

$$\frac{dP(0)}{dr} = 0 \qquad\qquad ; \text{ flat at origin}$$

As described in the text, one can simply choose the starting conditions in

---

[1]This equation of state is as stiff as possible, while still consistent with causal propagation of sound signals — see text.

Eqs. (10.79)

$$\rho(0) = \rho_0 \qquad\qquad ; \text{ finite}$$
$$P(\rho_0) = P_0$$

and integrate out on r until the pressure vanishes. This will give the total mass of the star for various ρ_0.

(b) As instructed, we use the stiff equation of state $P = \rho c^2$ over the range of densities relevant to neutron stars. This EOS is inappropriate in the low-density tail of the mass distribution, where the neutron matter is almost self-bound and the pressure vanishes. From Fig. 10.11 in the text, we thus model the EOS with

$$P = \rho c^2 \qquad ; \rho > \rho_{\min} \qquad ; \rho_{\min} \approx 10^{14}\,\text{gm/cm}^3$$
$$= 0 \qquad\quad ; \rho < \rho_{\min}$$

The dimensions can now be taken out of the TOV equations with

$$\tilde{r} \equiv \frac{r}{R_\odot^s} \qquad\qquad ; \tilde{\rho} \equiv \frac{(R_\odot^s)^3}{M_\odot}\,\rho$$

where $(M_\odot, R_\odot^s)$ are the mass and Schwarzschild radius of the sun [see Eqs. (10.87)–(10.88)]

$$R_\odot^s = \frac{2M_\odot G}{c^2} = 2.95 \times 10^5\,\text{cm} \qquad\qquad ; M_\odot = 1.99 \times 10^{33}\,\text{gm}$$

The dimensionless form of the TOV Eqs. (10.76) is then

$$\frac{d\tilde{\rho}(\tilde{r})}{d\tilde{r}} = -\frac{\tilde{M}(\tilde{r})\tilde{\rho}(\tilde{r})}{\tilde{r}^2}\left[\frac{1}{1 - \tilde{M}(\tilde{r})/\tilde{r}}\right]\left[1 + \frac{4\pi\tilde{r}^3\tilde{\rho}(\tilde{r})}{\tilde{M}(\tilde{r})}\right] \qquad ; \tilde{\rho} > \tilde{\rho}_{\min}$$

$$\tilde{M}(\tilde{r}) = \int_0^{\tilde{r}} 4\pi\tilde{\rho}\,\tilde{s}^2 d\tilde{s}$$

This first-order, non-linear, integro-differential equation can be converted into a finite-difference equation for the three-component column vector

$$\underline{Z} = \begin{bmatrix} \tilde{r} \\ \tilde{M}(\tilde{r}) \\ \tilde{\rho}(\tilde{r}) \end{bmatrix}$$

It reads

$$\underline{Z}_{n+1} = \underline{Z}_n + \Delta \times$$

$$\left[\frac{1}{4\pi(Z_n)_1^2(Z_n)_3} - \left[(Z_n)_2(Z_n)_3/(Z_n)_1^2\right]\left\{1/[1-(Z_n)_2/(Z_n)_1]\right\}\left[1+4\pi(Z_n)_1^3(Z_n)_3/(Z_n)_2\right] \right]$$

This equation was iterated using Mathcad 7, with starting values[2]

$$\underline{Z}_1 = \left[\begin{array}{c} \Delta \\ 4\pi\Delta^3\tilde{\rho}_0/3 \\ \tilde{\rho}_0 \end{array} \right]$$

The radius of the neutron star $\tilde{R}$ is that distance for which the starting density $\tilde{\rho}_0 = \tilde{\rho}(0)$ falls to $\tilde{\rho}_{min}$. The quantity $\tilde{M}(\tilde{R})$ is then the mass of the neutron star in units of $M_\odot$.

In Table 10.2 we give several results obtained from the above expresions starting from various values of $\tilde{\rho}_0$. Figure 10.1 shows $\rho(r)/\rho_0$ vs. r/R corresponding to the last line in that table.

Table 10.2   Some numerical results for the TOV equations with the EOS of Prob. 10.3(b).

$\tilde{\rho}_0$	$\tilde{R}$	ρ_0 (gm/cm^3)	R (cm)	$\rho(R)$(gm/cm^3)	$M/M_\odot$
1	5.47	7.76×10^{16}	1.61×10^6	1.002×10^{14}	2.70
0.5	5.40	3.88×10^{16}	1.59×10^6	1.003×10^{14}	2.69
0.1	5.47	7.76×10^{15}	1.61×10^6	1.002×10^{14}	2.91
0.05	5.68	3.88×10^{15}	1.68×10^6	1.002×10^{14}	3.08
0.01	6.53	7.76×10^{14}	1.93×10^6	1.000×10^{14}	3.23

(c) We make the following observations:

- It is clear from Fig. 10.11 in the text that the model EOS used here for neutron matter is as stiff as possible and *overestimates* the pressure in the region of mass densities relevant to neutron stars;
- In line with Fig. 10.11 in the text, the pressure is here simply defined to be zero when the mass density falls below $\rho_{min} = 10^{14}$ gm/cm^3;
- While expressing the overall behavior, this EOS thus mutilates the surface, and the more realistic EOS obtained from RMFT in

---

[2] We start from $n = 1$ in order to avoid any potential problems as $\tilde{r} \to 0$. We also use $\tilde{r}_n = n\tilde{R}/N$ and ensure $\Delta = \tilde{R}/N \to 0$. The Mathcad matrix notation was employed.

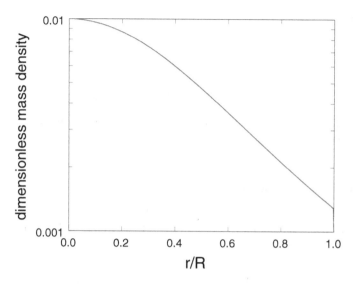

Fig. 10.1 Dimensionless mass density $\tilde{\rho}$ as a function of $\tilde{r}/\tilde{R} = r/R$ for the solution corresponding to the last row of Table 10.2, where $\tilde{\rho}_0 = 0.01$

Fig. 10.11 in the text should provide a continuous transition to zero density in Fig. 10.1;[3]

- While an overestimate, the behavior of the obtained values of $M/M_\odot$ indeed mirrors that in Fig. 10.12 in the text (see the discussion there). Here one finds

$$\left(\frac{M}{M_\odot}\right)_{\text{max}} \approx 3.23 \qquad ; \text{model EOS}$$

This should be compared with the value in Eq. (10.108) obtained with the more realistic equation of state from RMFT

$$\left(\frac{M}{M_\odot}\right)_{\text{max}} = 2.57 \qquad ; \text{RMFT}$$

Problem 10.4 (a) Derive the parametric equation of state of an ideal non-relativistic degenerate Fermi gas as a limiting form of Eqs. (10.98) and

---

[3]Although we have not demonstrated that in the solution to this problem. [See, however, Prob. 10.4(b).]

$(10.102)^4$

$$\rho c^2 = m_b c^2 \rho_{\rm B} + \frac{3}{5} \frac{\hbar^2}{2m_b} \left(\frac{6\pi^2}{\gamma}\right)^{2/3} \rho_{\rm B}^{5/3}$$

$$P = \frac{2}{3}(\rho c^2 - m_b c^2 \rho_{\rm B})$$

(b) Repeat Prob. 10.3 with this equation of state for neutron matter.

(c) At what densities does this non-relativistic calculation become inconsistent?

Solution to Problem 10.4

(a) If it is an ideal Fermi gas, then there are no meson fields present, and $(\phi_0, V_0) = 0$. The expectation value of the hamiltonian in Eq. (10.98) then simplifies to

$$\varepsilon(\rho_{\rm B}) = \frac{\gamma}{(2\pi)^3} \int_0^{k_{\rm F}} (M_b^2 + k^2)^{1/2} d^3 k$$

For a non-relativistic gas, the square root can be expanded as

$$(M_b^2 + k^2)^{1/2} \approx M_b + \frac{k^2}{2M_b}$$

Then

$$\frac{\gamma}{(2\pi)^3} \int_0^{k_{\rm F}} d^3 k = \frac{\gamma k_{\rm F}^3}{6\pi^2} = \rho_{\rm B}$$

$$\frac{\gamma}{(2\pi)^3} \int_0^{k_{\rm F}} \frac{k^2}{2M_b} d^3 k = \frac{3}{5} \frac{k_{\rm F}^2}{2M_b} \rho_{\rm B}$$

It follows that

$$\varepsilon(\rho_{\rm B}) \approx M_b \rho_{\rm B} + \frac{3}{5} \frac{1}{2M_b} \left(\frac{6\pi^2}{\gamma}\right)^{2/3} \rho_{\rm B}^{5/3}$$

The pressure follows from Eq. (10.101)

$$\mathcal{P} = \rho_{\rm B}^2 \frac{\partial}{\partial \rho_{\rm B}} \left(\frac{\varepsilon}{\rho_{\rm B}}\right)$$

$$= \frac{2}{5} \frac{1}{2M_b} \left(\frac{6\pi^2}{\gamma}\right)^{2/3} \rho_{\rm B}^{5/3}$$

---

[4]See chap. 2 of [Fetter and Walecka (2003a)] regarding the equations of state used in Probs. 10.4 and 10.5; recall that ρc^2 is the proper energy density.

Recall this analysis is in units where $\hbar = c = 1$, and everything is in inverse lengths.[5] To restore the correct units, let $\hbar c \varepsilon \to \rho c^2$, $\hbar c P \to P$, and $M_b \to m_b c/\hbar$ [see Eqs. (10.103)]. This gives

$$\rho c^2 = m_b c^2 \rho_B + \frac{3}{5} \frac{\hbar^2}{2m_b} \left(\frac{6\pi^2}{\gamma}\right)^{2/3} \rho_B^{5/3}$$

$$P = \frac{2}{5} \frac{\hbar^2}{2m_b} \left(\frac{6\pi^2}{\gamma}\right)^{2/3} \rho_B^{5/3}$$

which are the stated answers.

(b) First, we re-write the above EOS for neutron matter as

$$\rho = m_b \rho_B + \frac{3}{2} \frac{P}{c^2}$$

Now put this into the dimensionless units of Prob. 10.3

$$\tilde{\rho} = \frac{(R_\odot^s)^3}{M_\odot}(m_b \rho_B) + \frac{3}{2}\tilde{P} \qquad ; \tilde{P} = \frac{(R_\odot^s)^3}{M_\odot c^2}P$$

After inversion of the equation for P, and with $\gamma = 2$ for neutron matter, one finds

$$\frac{(R_\odot^s)^3}{M_\odot}(m_b \rho_B) = (R_\odot^s)^3 \frac{m_b}{M_\odot}\left[\frac{5}{2}\left(\frac{2m_b}{\hbar^2}\right)\right]^{3/5} \frac{1}{(3\pi^2)^{2/5}}\left[\frac{M_\odot c^2}{(R_\odot^s)^3}\right]^{3/5} \tilde{P}^{3/5}$$
$$\equiv \eta \tilde{P}^{3/5}$$

where the last line defines the dimensionless constant η. A combination of factors then gives

$$\eta = \left[\frac{(R_\odot^s)^3}{M_\odot}\frac{m_b}{(\hbar/m_b c)^3}\frac{5^{3/2}}{(3\pi^2)}\right]^{2/5}$$

Numerical evaluation with

$$M_\odot = 1.99 \times 10^{33}\,\text{gm} \qquad ; R_\odot^s = 2.95 \times 10^5\,\text{cm}$$

$$m_b = 1.67 \times 10^{-24}\,\text{gm} \qquad ; \frac{\hbar}{m_b c} = 2.10 \times 10^{-14}\,\text{cm}$$

gives η the following value[6]

$$\eta = 0.949$$

---

[5] See footnote 13 on p. 202 of the text.
[6] A remarkable result, considering the disparity of the input!

The appropriate EOS in these dimensionless units for the model of neutron matter as a non-interacting, non-relativistic Fermi gas is therefore

$$\tilde{\rho}(\tilde{P}) = \eta \tilde{P}^{3/5} + \frac{3}{2}\tilde{P} \qquad \text{; Fermi gas in NRL}$$

The corresponding TOV equation in these same dimensionless units is

$$\frac{d\tilde{P}(\tilde{r})}{d\tilde{r}} = -\frac{\tilde{M}(\tilde{r})\tilde{\rho}(\tilde{r})}{2\tilde{r}^2} \left[\frac{1 + \tilde{P}(\tilde{r})/\tilde{\rho}(\tilde{r})}{1 - \tilde{M}(\tilde{r})/\tilde{r}} \right] \left[1 + \frac{4\pi\tilde{r}^3\tilde{P}(\tilde{r})}{\tilde{M}(\tilde{r})} \right] \qquad \text{; TOV eqn}$$

$$\tilde{M}(\tilde{r}) = \int_0^{\tilde{r}} 4\pi\tilde{\rho}\,\tilde{s}^2 d\tilde{s}$$

where the local equation of state is now $\tilde{\rho}(\tilde{r}) = \tilde{\rho}[\tilde{P}(\tilde{r})]$.

In order to solve this first-order, non-linear, integro-differential equation, we introduce the three-component column vector

$$\underline{Z} = \begin{bmatrix} \tilde{r} \\ \tilde{M}(\tilde{r}) \\ \tilde{P}(\tilde{r}) \end{bmatrix}$$

and write the following finite-difference equation

$$\underline{Z}_{n+1} = \underline{Z}_n + \Delta \begin{bmatrix} 1 \\ 4\pi(Z_n)_1^2\,\tilde{\rho}[(Z_n)_3] \\ D_n \end{bmatrix}$$

where

$$D_n \equiv -\left[\frac{(Z_n)_2\,\tilde{\rho}[(Z_n)_3]}{2(Z_n)_1^2} \right] \left[\frac{1 + (Z_n)_3/\rho[(Z_n)_3]}{1 - (Z_n)_2/(Z_n)_1} \right] \left[1 + \frac{4\pi(Z_n)_1^3(Z_n)_3}{(Z_n)_2} \right]$$

This equation can again be iterated with Mathcad 7.

One solution is shown in Fig. 10.2.[7] In the dimensionless units, this example has

$$\tilde{P}(0) = 1 \qquad ; \frac{M}{M_\odot} = 0.468$$

$$\tilde{\rho}(0) = 2.449 \qquad ; \tilde{R} = 1.533$$

The corresponding dimensional quantities are

$$\rho(0) = 1.899 \times 10^{17}\,\frac{\text{gm}}{\text{cm}^3} \qquad ; R = 4.522 \times 10^5\,\text{cm}$$

---

[7]Here $\Delta = 10^{-3}$.

ONE SOLUTION TO TOV EQUATIONS WITH FERMI GAS IN NRL

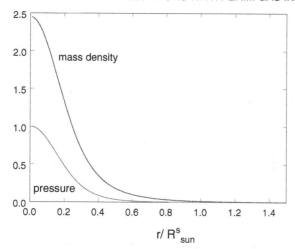

Fig. 10.2   One solution to the TOV equations for a neutron star with the EOS of a Fermi gas in the NRL. The pressure $\tilde{P}(\tilde{r})$ and mass density $\tilde{\rho}(\tilde{r})$ are shown in dimensionless units. Here $\tilde{P}(0) = 1$, $\tilde{\rho}(0) = 2.449$, $\tilde{R} = 1.533$, and $M/M_\odot = 0.468$.

A few comments:

- Here there is a well-defined radius where $P(R) = \rho(R) = 0$;[8]
- The surface, however, has an unrealistic long-range, low-density tail;
- A similar tail is observed in the comparable Thomas-Fermi theory of atomic structure (see [Walecka (2008)]);
- Although the central density is higher, the radius and mass are substantially smaller than any of those stars in Table 10.2, calculated with the model "stiff" EOS in Prob. 10.3;
- The estimate in the following part (c) shows the NRL is already inconsistent at the central density of this example.
- Table 10.3 shows results for some stars with a lower central density, within the limit in part (c);
- Figure 10.3 then shows the dimensionless pressure and density corresponding to the third line in Table 10.3, which has the maximum

---

[8]See the discussion in Prob. 10.3.

mass of

$$\left(\frac{M}{M_\odot}\right)_{\max} = 0.796$$

Table 10.3   Some numerical results for the TOV equations with the EOS of Prob. 10.4(a).

$\tilde{P}_0$	$\tilde{\rho}_0$	$\tilde{R}$	ρ_0 (gm/cm^3)	R (cm)	$M/M_\odot$
1	2.449	1.533	1.899×10^{17}	4.522×10^5	0.468
10^{-1}	0.388	1.896	3.011×10^{16}	5.593×10^5	0.662
10^{-2}	0.0749	2.825	5.806×10^{15}	8.334×10^5	0.796
10^{-3}	0.0165	4.133	1.283×10^{15}	1.219×10^6	0.704

A SECOND SOLUTION TO TOV EQUATIONS WITH FERMI GAS IN NRL

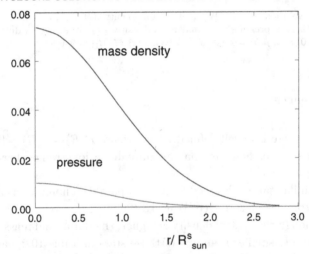

Fig. 10.3   A second solution to the TOV equations for a neutron star with the EOS of a Fermi gas in the NRL. The pressure $\tilde{P}(\tilde{r})$ and mass density $\tilde{\rho}(\tilde{r})$ are shown in dimensionless units. Here $\tilde{P}(0) = 0.01$, $\tilde{\rho}(0) = 0.0749$, $\tilde{R} = 2.825$, and $M/M_\odot = 0.796$.

(c) Let's say the NRL calculation becomes inconsistent when the Fermi energy exceeds 1/2 the baryon rest mass (a generous estimate)

$$\frac{\hbar^2 k_F^2}{2m_b} \approx \frac{1}{2} m_b c^2$$

This is re-written in terms of baryon density as

$$\rho_B \approx \frac{1}{3\pi^2} \left(\frac{m_b c}{\hbar}\right)^3 = 3.64 \times 10^{39} \, \text{cm}^{-3}$$

or, in terms of a mass density,

$$m_b \rho_B \approx 6.09 \times 10^{15} \, \text{gm/cm}^3$$

It is clear from Fig. 10.11 in the text that the mass densities appropriate to neutron stars already approach this value.[9]

Problem 10.5 Repeat Prob. 10.3 with the equation of state of an ideal ultra-relativistic gas[10]

$$P = \frac{1}{3}\rho c^2$$

Compare with the results in Prob. 10.3.

Solution to Problem 10.5

If one neglects the meson fields and baryon mass in the RMFT equation of state in Eqs. (10.98) and (10.102), they become

$$\mathcal{P} = \frac{1}{3} \frac{\gamma}{(2\pi)^3} \int^{k_F} k \, d^3 k = \frac{1}{3}\varepsilon$$

When the units are restored according to Eqs. (10.103), this gives the EOS of an ideal ultra-relativistic gas

$$P = \frac{1}{3}\rho c^2$$

As instructed, we repeat Prob. 10.3 with the substitution $\rho c^2 \to \rho c^2/3$ in the EOS for dense neutron matter. The surface is treated exactly as before. The dimensionless form of the TOV equations is now

$$\frac{d\tilde{\rho}(\tilde{r})}{d\tilde{r}} = -\frac{2\tilde{M}(\tilde{r})\tilde{\rho}(\tilde{r})}{\tilde{r}^2} \left[\frac{1}{1 - \tilde{M}(\tilde{r})/\tilde{r}}\right] \left[1 + \frac{4\pi\tilde{r}^3\tilde{\rho}(\tilde{r})}{3\tilde{M}(\tilde{r})}\right] \qquad ; \tilde{\rho} > \tilde{\rho}_{\text{min}}$$

$$\tilde{M}(\tilde{r}) = \int_0^{\tilde{r}} 4\pi\tilde{\rho}\,\tilde{s}^2 d\tilde{s}$$

---

[9] And the first solution in part (b) exceeds it!

[10] This is the equation of state of a high-pressure, asymptotically-free, collection of massless quarks and gluons (see [Walecka (2004)]).

Table 10.4 shows some numerical results, again obtained from the finite-difference equation using Mathcad 7. These results should be compared with those in Table 10.2. The radii and masses are somewhat smaller in this case; however, the trends are exactly the same. Here one finds

$$\left(\frac{M}{M_\odot}\right)_{\max} \approx 2.87$$

which represents a reduction of $\approx 10\%$ from the model EOS result in Prob. 10.3.

Table 10.4   Some numerical results for the TOV equations with the EOS for dense neutron matter in Prob. 10.3(b) modified from $P = \rho c^2$ to $P = \rho c^2/3$.

$\tilde{\rho}_0$	$\tilde{R}$	ρ_0 (gm/cm^3)	R (cm)	$\rho(R)$(gm/cm^3)	$M/M_\odot$
1	4.99	7.76×10^{16}	1.47×10^6	1.000×10^{14}	2.05
0.5	4.86	3.88×10^{16}	1.43×10^6	1.003×10^{14}	2.04
0.1	4.93	7.76×10^{15}	1.45×10^6	1.002×10^{14}	2.31
0.05	5.20	3.88×10^{15}	1.53×10^6	1.002×10^{14}	2.53
0.01	6.22	7.76×10^{14}	1.83×10^6	1.000×10^{14}	2.87

Problem 10.6 Use the first law of thermodynamics in Eq. (10.101) to derive the RMFT expression for the pressure in Eq. (10.102).

Solution to Problem 10.6

The RMFT expression for the energy density is given in Eq. (10.98)

$$\varepsilon(\rho_B, M_b^\star) = \frac{1}{2}\frac{g_v^2}{m_v^2}\rho_B^2 + \frac{1}{2}\frac{m_s^2}{g_s^2}(M_b - M_b^\star)^2 + \frac{\gamma}{(2\pi)^3}\int_0^{k_F}(k^2 + M_b^{\star 2})^{1/2}\, d^3k$$

where we now denote this by $\varepsilon(\rho_B, M_b^\star)$ [see Eq. (10.95)].

The pressure follows from the first law of thermodynamics applied to this degenerate Fermi gas [see Eq. (10.101)]

$$\mathcal{P} = \rho_B^2 \frac{\partial}{\partial \rho_B}\left(\frac{\varepsilon}{\rho_B}\right)$$

Recall that this derivative is to be carried out at fixed B.

We first show that

$$\frac{\partial \varepsilon(\rho_B, M_b^\star)}{\partial M_b^\star} = 0$$

where the partial here means the other member of the set $(\rho_{\rm B}, M_b^\star)$ is to be held fixed. This relation implies that one can keep $M_b^\star$ *fixed* in computing the pressure. It is obtained by just carrying out the derivative

$$\frac{\partial \varepsilon(\rho_{\rm B}, M_b^\star)}{\partial M_b^\star} = -\frac{m_s^2}{g_s^2}(M_b - M_b^\star) + \frac{\gamma}{(2\pi)^3} \int_0^{k_{\rm F}} \frac{M_b^\star}{(k^2 + M_b^{\star 2})^{1/2}} d^3k$$

However, this result *vanishes* by the self-consistency Eq. (10.100)

$$\frac{\partial \varepsilon(\rho_{\rm B}, M_b^\star)}{\partial M_b^\star} = 0 \qquad ; \text{ self-consistency}$$

The pressure is then given by

$$\mathcal{P} = \frac{1}{2}\frac{g_v^2}{m_v^2}\rho_{\rm B}^2 - \frac{1}{2}\frac{m_s^2}{g_s^2}(M_b - M_b^\star)^2 - \frac{\gamma}{(2\pi)^3}\int_0^{k_{\rm F}}(k^2 + M_b^{\star 2})^{1/2}d^3k$$
$$+\rho_{\rm B}\frac{\gamma}{(2\pi)^3}4\pi k_{\rm F}^2(k_{\rm F}^2 + M_b^{\star 2})^{1/2}\frac{dk_{\rm F}}{d\rho_{\rm B}}$$

where the last term arises from differentiating the integral with respect to its upper limit. With $d\rho_{\rm B}/dk_{\rm F} = \gamma k_{\rm F}^2/2\pi^2$, this becomes

$$\mathcal{P} = \frac{1}{2}\frac{g_v^2}{m_v^2}\rho_{\rm B}^2 - \frac{1}{2}\frac{m_s^2}{g_s^2}(M_b - M_b^\star)^2 - \frac{\gamma}{(2\pi)^3}\int_0^{k_{\rm F}}(k^2 + M_b^{\star 2})^{1/2}d^3k$$
$$+\rho_{\rm B}(k_{\rm F}^2 + M_b^{\star 2})^{1/2}$$

Now consider the integral

$$\mathcal{I} \equiv \frac{1}{3}\frac{\gamma}{(2\pi)^3}\int_0^{k_{\rm F}}\frac{k^2}{(k^2 + M_b^{\star 2})^{1/2}}d^3k$$
$$= \frac{\gamma}{6\pi^2}\int_0^{k_{\rm F}}\frac{k^4 dk}{(k^2 + M_b^{\star 2})^{1/2}}$$

Carry out a partial integration

$$du = \frac{k dk}{(k^2 + M_b^{\star 2})^{1/2}} \qquad ; v = k^3$$
$$u = (k^2 + M_b^{\star 2})^{1/2} \qquad ; dv = 3k^2 dk$$

Then

$$\mathcal{I} = \rho_{\rm B}(k_{\rm F}^2 + M_b^{\star 2})^{1/2} - \frac{\gamma}{(2\pi)^3}\int_0^{k_{\rm F}}(k^2 + M_b^{\star 2})^{1/2}d^3k$$

Hence

$$P = \frac{1}{2}\frac{g_v^2}{m_v^2}\rho_B^2 - \frac{1}{2}\frac{m_s^2}{g_s^2}(M_b - M_b^\star)^2 + \frac{1}{3}\frac{\gamma}{(2\pi)^3}\int_0^{k_F}\frac{k^2}{(k^2 + M_b^{\star 2})^{1/2}}d^3k$$

This is Eq. (10.102).

Problem 10.7 Show that with the equation of state $P = \rho c^2$, the thermo-dynamic speed of sound is equal to the speed of light. (*Hint:* see chap. 49 of [Fetter and Walecka (2003)]; note that there $c^2 \equiv c_{\text{sound}}^2$.)

Solution to Problem 10.7

Consider a uniform medium with a mass density $\rho(\vec{x}, t)$, hydrodynamic velocity $\vec{v}(\vec{x}, t)$, and pressure $P(\vec{x}, t)$. The continuity equation and Newton's second law give

$$\frac{\partial \rho}{\partial t} + \vec{\nabla} \cdot (\rho \vec{v}) = 0$$

$$\frac{d}{dt}(\rho \vec{v}) = -\vec{\nabla}P$$

The substantive, or total, time derivative of the velocity field is

$$\frac{d\vec{v}}{dt} = \frac{\partial \vec{v}}{\partial t} + (\vec{v} \cdot \vec{\nabla})\vec{v}$$

Suppose one looks for a solution with $\rho = \rho_0 + \delta\rho$, and linearizes in $(\delta\rho, \vec{v})$. The first two equations above then become

$$\frac{\partial \, \delta\rho}{\partial t} + \vec{\nabla} \cdot (\rho_0 \vec{v}) = 0$$

$$\frac{\partial(\rho_0 \vec{v})}{\partial t} = -\vec{\nabla}P$$

The partial time derivative of the first equation, combined with the divergence of the second, gives

$$\frac{\partial^2 \delta\rho}{\partial t^2} = \nabla^2 P$$

We now introduce the equation of state $P(\rho)$, and expand to first order

$$P(\rho_0 + \delta\rho) = P(\rho_0) + \left(\frac{dP}{d\rho}\right)_{\rho_0} \delta\rho$$

Then

$$\frac{\partial^2 \delta\rho}{\partial t^2} = c_s^2 \nabla^2 \delta\rho \qquad ; \ c_s^2 \equiv \left(\frac{dP}{d\rho}\right)_{\rho_0}$$

This is the wave equation for hydrodynamic sound waves with sound velocity c_s.

The RMFT gives the parametric equation of state in Eqs. (10.98) and (10.102). At high baryon density, these yield

$$\varepsilon = \frac{1}{2}\frac{g_v^2}{m_v^2}\rho_B^2 \qquad ; \ \rho_B \to \infty$$

$$\mathcal{P} = \frac{1}{2}\frac{g_v^2}{m_v^2}\rho_B^2$$

Thus at high baryon density, with the aid of Eqs. (10.103), RMFT gives the equation of state

$$P = \rho c^2 \qquad ; \ \rho_B \to \infty$$

In this limit, the speed of hydrodynamic sound is equal to the velocity of light[11]

$$c_s^2 = \frac{dP}{d\rho} = c^2$$

Problem 10.8 Consider the (p, e^-) component present at equilibrium in a cold, neutral neutron star in RMFT. Assume massless electrons.

(a) Show the Fermi wavenumber of the protons is related to that of the neutrons by

$$k_{Fp} + \left(k_{Fp}^2 + M_b^{\star\,2}\right)^{1/2} = \left(k_{Fn}^2 + M_b^{\star\,2}\right)^{1/2}$$

(b) What is the ratio of proton to neutron densities as $M_b^\star \to \infty$? As $M_b^\star \to 0$?

(c) Estimate from Figs. 10.10 and 10.11 in the text the ratio of proton to neutron densities actually present in the neutron star.

Solution to Problem 10.8

(a) We are investigating the (p, e^-) component present at equilibrium in a cold, neutral neutron star in RMFT, assuming massless electrons. For

---

[11]It is actually *zero sound* that propagates in this degenerate Fermi system [Fetter and Walecka (2003a)].

electrical neutrality, the electron density must be equal to the proton density. From Eq. (10.96), this implies

$$\rho_e = \rho_p = \frac{k_{\mathrm{Fp}}^3}{3\pi^2} \qquad \text{; neutrality}$$

The neutron density is

$$\rho_n = \frac{k_{\mathrm{Fn}}^3}{3\pi^2} \qquad \text{; neutrons}$$

For stability, the following relation must be satisfied by the Fermi energies (we assume massless electrons)

$$k_{\mathrm{Fp}} + \left(k_{\mathrm{Fp}}^2 + M_b^{\star\,2}\right)^{1/2} = \left(k_{\mathrm{Fn}}^2 + M_b^{\star\,2}\right)^{1/2} \qquad \text{; stability}$$

- If the l.h.s. were larger, then the energy of the system could be lowered through the reaction

$$e^- + p \rightarrow n + \nu_e$$

where ν_e is a low-energy, massless, electron neutrino;
- If the r.h.s. were larger, then the energy of the system could be lowered through the reaction

$$n \rightarrow p + e^- + \bar{\nu}_e$$

where $\bar{\nu}_e$ is a low-energy, massless, electron antineutrino.

(b) As $M_b^\star \rightarrow \infty$, the above relation becomes

$$k_{\mathrm{Fp}} = 0 \qquad \text{; } M_b^\star \rightarrow \infty$$

In this limit, there is *no* (p, e^-) component in the neutron star

$$\rho_e = \rho_p = 0$$

As $M_b^\star \rightarrow 0$, the above relation becomes

$$2k_{\mathrm{Fp}} = k_{\mathrm{Fn}} \qquad \text{; } M_b^\star \rightarrow 0$$

In this limit, the densities in the neutron star satisfy

$$\rho_e = \rho_p = \frac{1}{8}\rho_n$$

(c) With RMFT in the high-density regime where $\mathcal{P} = \varepsilon$, one has from Eqs. (10.98), (10.102), and (10.105)

$$\mathcal{P} = \varepsilon = \frac{1}{2}\frac{g_v^2}{m_v^2}\rho_B^2$$

$$= \frac{1}{2}C_v^2\frac{\rho_B^2}{M_b^2} \qquad ; \ C_v^2 = 195.9$$

The units are restored through Eqs. (10.103)

$$\rho c^2 = \hbar c \,\varepsilon \qquad ; \ \text{energy density}$$

$$P = \hbar c\,\mathcal{P} \qquad \text{pressure}$$

$$M_b = \frac{m_b c}{\hbar} \qquad \text{nucleon mass}$$

Thus, in the high-density regime

$$P = \rho c^2 = \frac{\hbar c}{2}C_v^2\frac{\rho_B^2}{M_b^2}$$

Now go to Fig. 10.11 in the text, and take for the neutron star[12]

$$P \approx \rho c^2 \approx 10^{36}\ \frac{\text{gm}}{\text{cm-sec}^2}$$

In c.g.s. units

$$\hbar = 1.055 \times 10^{-27}\,\text{erg-sec} \qquad ; \ c = 2.998 \times 10^{10}\,\text{cm/sec}$$
$$M_b^{-1} = 2.103 \times 10^{-14}\,\text{cm}$$

This gives, from the above,

$$\rho_B \approx 0.854\ \text{F}^{-3} \qquad ; \ 1\,\text{F} = 10^{-13}\,\text{cm}$$

With $\rho_B \approx \rho_n = k_{Fn}^3/3\pi^2$, this implies

$$k_{Fn} \approx 2.93\ \text{F}^{-1}$$

Now go to Fig. 10.10 in the text. At this value of k_{Fn}, one has in neutron matter

$$\frac{M_b^\star}{M_b} \approx 0.20$$

The solution to the Fermi energy relation in part (a) then gives

$$k_{Fp} \approx 1.39\ \text{F}^{-1}$$

---

[12] Recall footnote 21 on p. 209.

Hence the ratio of proton to neutron densities actually present in this neutron star is

$$\frac{\rho_p}{\rho_n} = \left(\frac{k_{\mathrm{Fp}}}{k_{\mathrm{Fn}}}\right)^3 \approx 10\%$$

Problem 10.9 Problems 10.9–10.12 concern the *thermodynamics* of black holes. We make no attempt to derive these relations from first principles. They are included only so that the reader has some acquaintance with these concepts when approaching the literature.[13]

A black hole can be assigned a *temperature*

$$k_{\mathrm{B}}T = \frac{\hbar}{2\pi c}g_s$$

Here $k_{\mathrm{B}} = 1.381 \times 10^{-16}\,\mathrm{erg}/^\circ\mathrm{K}$ is Boltzmann's constant, and g_s is the acceleration of gravity at the Schwarzschild radius

$$g_s = \frac{MG}{R_s^2}$$

(a) Show that the relation for the temperature reduces to[14]

$$k_{\mathrm{B}}T = \frac{\hbar c^3}{8\pi GM}$$

Note the $\hbar$ in this relation — it is a quantum effect. Note also the characteristic energy $(\hbar c^3/GM)$.

(b) Use the numbers in Eqs. (10.87), and $\hbar = h/2\pi = 1.055 \times 10^{-27}\,\mathrm{erg\text{-}sec}$, to compute the temperature of a black hole with mass equal to the solar mass $M = M_\odot$.

Solution to Problem 10.9

(a) We are given that the assigned temperature of a black hole is

$$k_{\mathrm{B}}T = \frac{\hbar}{2\pi c}g_s$$

---

[13]See, for example, [Ohanian and Ruffini (1994)], [Taylor and Wheeler (2000)], [Weinberg (1972)].

[14]This, and many other topics, are conveniently referenced on the internet at [Wiki (2006)]. See also [Batiz (2000)].

where g_s is the acceleration of gravity at the Schwarzschild radius $R_s = 2MG/c^2$

$$g_s = \frac{MG}{R_s^2} \qquad ; \; R_s = \frac{2MG}{c^2}$$

A combination of these relations gives

$$k_B T = \frac{\hbar c^3}{8\pi G M}$$

(b) The numbers in Eqs. (10.87), as well as those given above, are in c.g.s.

$$c = 3.00 \times 10^{10}\,\text{cm/sec} \qquad ; \; k_B = 1.381 \times 10^{-16}\,\text{erg/}^\circ\text{K}$$
$$G = 6.67 \times 10^{-8}\,\text{cm}^3\text{/gm-sec}^2 \qquad ; \; \hbar = h/2\pi = 1.055 \times 10^{-27}\,\text{erg-sec}$$
$$M_\odot = 1.99 \times 10^{33}\,\text{gm}$$

These give

$$T = 6.18 \times 10^{-8}\,^\circ\text{K}$$

This is a very low temperature, indeed.

Problem 10.10 The Stefan-Boltzmann law gives the energy flux of electromagnetic radiation at the surface of a *black body* at a temperature T as [Ohanian (1995)]

$$\mathcal{S} = \sigma T^4 \qquad ; \text{Stefan-Boltzmann law}$$
$$\sigma = \frac{\pi^2 k_B^4}{60\hbar^3 c^2}$$

Note that the Stefan-Boltzmann constant σ does not depend on the emission or absorption *mechanism*, but only on $(k_B, \hbar, c)$.

A black hole loses energy by *Hawking radiation*—essentially pair production in the strong gravitational gradient at the Schwarzschild radius, where one member of the pair escapes.

(a) To estimate the lifetime of a black hole against "evaporation," assume that in the global inertial laboratory frame the rate of loss of mass due to Hawking radiation is given by

$$\frac{d}{dt}(Mc^2) = -(4\pi R_s^2)\mathcal{S}$$

Integrate this relation to find the time τ for disappearance of the black hole

$$\tau = \frac{1}{3} \frac{Mc^2}{(4\pi R_s^2)\mathcal{S}}$$

where the quantities on the r.h.s. are initial values.

(b) Rewrite this last result as

$$\tau = 5120\pi \left(\frac{G^2 M^3}{\hbar c^4} \right)$$

Note the dependence on the characteristic time $(G^2 M^3/\hbar c^4)$. [15]

(c) Evaluate τ (in years) for a black hole of one solar mass, $M = M_\odot$.

(d) Evaluate τ (in years) for a black hole of mass $M = 1\,\mathrm{kg}$.

(e) Repeat for $M = m_p$, where $m_p = 1.673 \times 10^{-24}$ gm is the mass of the proton.

Solution to Problem 10.10

(a) Both the Schwarzschild radius and temperature depend on the mass (see Prob. 10.9). If the mass dependence is scaled out and made explicit, the energy-loss equation is re-written as

$$\frac{dM}{dt} = - \left[\frac{(4\pi R_s^2)\mathcal{S}}{c^2} \right]_0 \frac{1}{M^2}$$

Now integrate on the time from $t = 0$ where the initial mass is M, until $t = \tau$, where the mass has disappeared

$$\int_0^\tau dt = \tau \qquad ; \quad -\int_M^0 M^2\, dM = \frac{1}{3}M^3$$

Hence the integral of the energy-loss equation takes the form

$$\frac{1}{3}M = \left[\frac{(4\pi R_s^2)\mathcal{S}}{c^2} \right]_0 \frac{\tau}{M^2}$$

where M is now the initial value. The lifetime of the black hole is therefore given by

$$\tau = \frac{1}{3} \frac{Mc^2}{(4\pi R_s^2)\mathcal{S}}$$

---

[15]The *Hawking lifetime* of a black hole is indeed given by $\tau_{\mathrm{H}} = 5120\pi G^2 M^3/\hbar c^4$ [Wiki (2006)].

(b) The result in (a) can be re-written in terms of the constants $(G, M, c, \hbar)$ using the expression for the temperature in Prob. 10.9

$$k_B T = \frac{\hbar c^3}{8\pi G M}$$

Thus

$$S = \frac{\pi^2}{60\hbar^3 c^2}\left(\frac{\hbar c^3}{8\pi G M}\right)^4$$

$$\tau = \frac{Mc^6}{12\pi(2MG)^2}\left(\frac{60\hbar^3 c^2}{\pi^2}\right)\left(\frac{8\pi M G}{\hbar c^3}\right)^4$$

A collection of factors then yields the Hawking lifetime of the black hole

$$\tau = 5120\pi\left(\frac{G^2 M^3}{\hbar c^4}\right)$$

(c) The lifetime of a black hole of one solar mass, $M = M_\odot$, follows from the numbers in Prob. 10.9(b)

$$\tau = 6.60 \times 10^{74}\,\mathrm{sec} = 2.09 \times 10^{67}\,\mathrm{yrs} \qquad ; M = M_\odot$$

This is interesting, but so many orders of magnitude larger than the age of the universe, apparently irrelevant.

(d) The lifetimes scales as M^3, and hence it is easy to get the other values. For $M = 1\,\mathrm{kg} = 10^3\,\mathrm{gm}$

$$\tau = 2.65 \times 10^{-24}\,\mathrm{yrs} \qquad ; M = 10^3\,\mathrm{gm}$$

(e) For a black hole with the proton mass, $M = m_p = 1.673 \times 10^{-24}\,\mathrm{gm}$

$$\tau = 1.24 \times 10^{-104}\,\mathrm{yrs} \qquad ; M = m_p = 1.673 \times 10^{-24}\,\mathrm{gm}$$

A black hole with the mass of a proton does not live very long!

Problem 10.11 An *entropy* can be associated with a black hole according to[16]

$$TS = \frac{1}{2}Mc^2$$

where T is the temperature introduced in Prob. 10.9(a).

---

[16]Recall that the Helmholtz free energy of a system is $F = E - TS$, so the product TS should scale like the energy, which in the case of a black hole is Mc^2. We leave it to the dedicated reader to supply an explanation for the factor of $1/2$.

(a) Show this relation can be rewritten as

$$\frac{S}{k_B} = \frac{4\pi M^2 G}{\hbar c} = \frac{c^3}{4\hbar G} A_s$$

where A_s is the *area* of a sphere with the Schwarzschild radius

$$A_s = 4\pi R_s^2 \qquad ; \text{ area}$$

(b) Discuss how the second law of thermodynamics might now be applied to black holes.

Solution to Problem 10.11

(a) From Prob. 10.9, the temperature of the black hole is

$$k_B T = \frac{\hbar c^3}{8\pi G M}$$

Hence the entropy introduced above is given by

$$\frac{S}{k_B} = \frac{Mc^2}{2k_B T} = \frac{4\pi M^2 G}{\hbar c}$$

The *area* of the black hole is defined by

$$A_s \equiv 4\pi R_s^2 \qquad ; \text{ area}$$

where the Schwarzschild radius is $R_s = 2MG/c^2$. Hence the entropy of the black hole can be written as[17]

$$\frac{S}{k_B} = \frac{c^3}{4\hbar G} A_s \qquad ; \text{ entropy}$$

(b) The two principal properties of the entropy which follow from the second law of thermodynamics are:

- The entropy of a closed system can only increase, $\delta S \geq 0$;
- The reversible heat flow to the system is $dQ_R = TdS$.

If the thermodynamic variable is A_s, then thermodynamics implies

$$\delta A_s \geq 0 \qquad\qquad ; \text{ closed system}$$
$$dQ_R = \frac{1}{2}Mc^2\frac{dA_s}{A_s} \qquad ; \text{ reversible heat flow}$$

---

[17]Since the area is an instantaneous, transverse quantity, the area the record keeper observes with the Schwarzschild metric, is the physical area.

Problem 10.12 Use the *equivalence principle* to generalize the result in Prob. 10.9 and associate a temperature (and acompanying Planck distribution of photons) with a particle undergoing an acceleration a_s according to

$$k_B T = \frac{\hbar}{2\pi c} a_s$$

It has been proposed that this will be an additional source of "Unruh" radiation for strongly accelerated electrons [Unruh (1976)], [Chen and Tajima (1999)].

Solution to Problem 10.12

The equivalence principle states that the inertial mass is identical to the gravitational mass. As a consequence, a local gravitational field is fully equivalent to a local accelerating coordinate system, and this leads to the concept of the local freely falling frame (LF^3).

In Prob. 10.9, a black hole is assigned a temperature through the relation

$$k_B T = \frac{\hbar}{2\pi c} g_s$$

Here k_B is Boltzmann's constant, and g_s is the acceleration of gravity at the Schwarzschild radius. The implications for Hawking radiation are explored in Prob. 10.10.

The equivalence principle would appear to imply that this relation can be generalized to read [Unruh (1976)], [Chen and Tajima (1999)]

$$k_B T = \frac{\hbar}{2\pi c} a_s$$

where a_s is now *any* acceleration, in particular that of charged, radiating electrons. These authors conclude that in addition to bremsstrahlung, there will be "Unruh" black-body radiation corresponding to a Planck distribution of photons at the temperature T.

To the best of the author's knowledge, this Unruh radiation has never been detected.

Chapter 11

Cosmology

Problem 11.1 Show that for the cosmology considered in this chapter, $\Lambda(t)$ and $\rho(t)$ develop in time according to

$$\Lambda^3(t) \propto (t - t_0)^2 \qquad ; \ \rho(t) \propto (t - t_0)^{-2}$$

Hence conclude that

$$\rho(t) \Lambda^3(t) = \text{constant}$$

Why might one have anticipated this result?

Solution to Problem 11.1

In this chapter, the Einstein field equations are solved for the Robertson-Walker metric with $k = 0$. Two consequences are those in Eqs. (11.36) and (11.39)

$$t - t_0 = \left[\frac{1}{6\pi G \rho(t)} \right]^{1/2} \qquad ; \ \text{age of universe}$$

$$\Lambda^2(t) = \gamma^{2/3}(t - t_0)^{4/3} \qquad ; \ \gamma = \text{constant}$$

It follows from these relations that

$$\Lambda^3(t) \propto (t - t_0)^2 \qquad ; \ \rho(t) \propto (t - t_0)^{-2}$$

We conclude that

$$\rho(t) \Lambda^3(t) = \text{constant}$$

Since the spatial volume is expanding as $\Lambda^3(t)$, we would anticipate that the mass density $\rho(t)$, arising, for example, from a given set of particles, should decrease as $1/\Lambda^3(t)$.

Problem 11.2 Do not assume $P \ll \rho c^2$, and retain the pressure P in the source terms in the Einstein field equations for the cosmology studied in this chapter.

(a) Show that Eqs. (11.27) are modified as follows

$$-\frac{3}{c^2}\frac{\ddot{\Lambda}}{\Lambda} = \frac{\kappa}{2}(\rho c^2 + 3P)$$

$$\frac{1}{c^2}\left[\frac{\ddot{\Lambda}}{\Lambda} + 2\left(\frac{\dot{\Lambda}}{\Lambda}\right)^2\right] = \frac{\kappa}{2}(\rho c^2 - P)$$

(b) Show that Eq. (11.34) becomes

$$h^2 = \left(\frac{\dot{\Lambda}}{\Lambda}\right)^2 = \frac{1}{3}\kappa\rho c^4$$

Thus this relation and its consequences are *independent* of P. What are those consequences?

Solution to Problem 11.2

(a) It follows from Eq. (10.1) that if the pressure is included, the energy-momentum tensor for a fluid at rest in the lab frame in Eq. (11.21) becomes

$$T^{\mu\nu} = Pg^{\mu\nu} + \left(\rho + \frac{P}{c^2}\right)u^\mu u^\nu$$

Equation (11.23) then becomes

$$T = T^\mu{}_\mu = 4P - (\rho c^2 + P) = 3P - \rho c^2$$

With the metric of Eqs. (11.8), and the four-velocity in Eqs. (11.22), the source terms for the Einstein field equations in Eqs. (11.24)–(11.25) take the form

$$T_{\mu\nu} - \frac{1}{2}T g_{\mu\nu} = \frac{1}{2}(\rho c^2 - P)g_{\mu\nu} + \left(\rho + \frac{P}{c^2}\right)u_\mu u_\nu$$

$$= \frac{1}{2}(\rho c^2 + 3P) \qquad ;\ \underline{44}$$

$$= 0 \qquad ;\ \underline{i4}$$

$$= \frac{\Lambda^2}{2}(\rho c^2 - P)\delta_{ij} \qquad ;\ \underline{ij}$$

It then follows from Eqs. (11.19) and (11.26) that with the inclusion of the pressure P, the field Eqs. (11.27) become

$$-\frac{3}{c^2}\frac{\ddot{\Lambda}}{\Lambda} = \frac{\kappa}{2}(\rho c^2 + 3P)$$

$$\frac{1}{c^2}\left[\frac{\ddot{\Lambda}}{\Lambda} + 2\left(\frac{\dot{\Lambda}}{\Lambda}\right)^2\right] = \frac{\kappa}{2}(\rho c^2 - P)$$

(b) Equations (11.30) then take the form

$$-3\left(\frac{dh}{dt} + h^2\right) = \frac{\kappa c^2}{2}(\rho c^2 + 3P)$$

$$\frac{dh}{dt} + 3h^2 = \frac{\kappa c^2}{2}(\rho c^2 - P)$$

Add 3/4 of the second equation to 1/4 of the first, to obtain

$$\frac{3}{2}h^2 = \frac{\kappa}{2}\rho c^4$$

This is Eq. (11.34), which therefore *remains valid* even when the pressure P is included in the source. It follows that

$$h^2 = \left(\frac{\dot{\Lambda}}{\Lambda}\right)^2 = \frac{1}{3}\kappa\rho c^4$$

As a consequence, the logarithmic time growth of the metric is directly related to the mass density, independent of P. In particular, Eq. (11.41) still determines Hubble's constant in terms of the mass density as[1]

$$H_0 \equiv \left[\frac{1}{\Lambda}\frac{d\Lambda}{dt}\right]_{t_p} = \left[\frac{1}{3}\kappa\rho(t_p)c^4\right]^{1/2}$$

Problem 11.3 The LF^3 in this cosmology is an unchanging spatially global frame in which there is neither gravity nor inertial forces. In the LF^3, one has only the laws of special relativity. The time t in the LF^3 is identical to that in the global inertial laboratory frame I, and at any instant in time, the LF^3 is tangent to the laboratory space.[2] The *proper distance*

---

[1] The observed baryon density is too small to explain the observed Hubble's constant. The difference is presumably made up by "dark matter".

[2] The LF^3 is global in space, but *local in time*.

is the distance at any instant in the LF^3. The coordinate transformation between the frames is given by Eq. (11.53).[3]

(a) Show that if the physical spatial distance in the global laboratory frame at any instant is l, then l is also the proper distance;

(b) Show that if a light signal is emitted at a time t from a star that is a proper distance l away from the origin O, then the time it takes to arrive at the origin in the LF^3 is $\Delta t = l/c$;

(c) Show that if the star is at a physical distance $l + \Delta l$ in the global inertial laboratory frame when the light signal arrives at the origin at the time $t + \Delta t$, then $l + \Delta l$ is the proper distance at that time.

Solution to Problem 11.3

(a) At a given instant in time in the LF^3, the finite interval is

$$\mathbf{s}^2 = \bar{l}^2$$

where $\bar{l}$ is the proper distance. This is also the physical interval in the global laboratory frame. Since the times in the two frames are identical, one has

$$\mathbf{s}^2 = l^2$$

where l is now the physical distance in the the global laboratory frame.[4] Hence, at any instant,

$$l = \bar{l} \qquad ; \text{ proper distance}$$

(b) Light follows the null interval. Hence if a light signal is emitted at a time t from a star that is a proper distance $\bar{l}$ away from the origin O, then the time Δt it takes to arrive at the origin in the LF^3 is given by

$$\mathbf{s}^2 = \bar{l}^2 - c^2(\Delta t)^2 = l^2 - c^2(\Delta t)^2 = 0$$

---

[3]The transformation matrix that takes us from the local freely falling frame LF^3 with coordinates $\bar{q}^\mu = (\bar{x}, \bar{y}, \bar{z}, c\bar{t})$, where there is no gravity and only special relativity, to the global inertial laboratory frame I with coordinates $q^\mu = (q^1, q^2, q^3, ct)$, is given in Eqs. (11.53) as

$$\underline{a} = \begin{bmatrix} 1/\Lambda(t) & 0 & 0 & 0 \\ 0 & 1/\Lambda(t) & 0 & 0 \\ 0 & 0 & 1/\Lambda(t) & 0 \\ 0 & 0 & 0 & 1 \end{bmatrix}$$

The time is unmodified by the uniform mass distribution, while the spatial coordinates are instantaneously stretching.

[4]Note this is distinct from the *coordinate* distance.

Hence

$$\Delta t = \frac{l}{c}$$

(c) Suppose the star in part (b) is at a physical distance $l + \Delta l$ in the global inertial laboratory frame when the light signal arrives at the origin at the time $t + \Delta t$. The interval in the LF^3 from the time the light signal was emitted is

$$(\Delta s)^2 = (\Delta \bar{l})^2 - c^2(\Delta t)^2$$

This same interval in the laboratory frame is

$$(\Delta s)^2 = (\Delta l)^2 - c^2(\Delta t)^2$$

Hence

$$\Delta l = \Delta \bar{l} \qquad ; \text{ proper distance}$$

Thus the physical distance Δl that the star has traveled in the global inertial laboratory frame in the time inteval Δt is also the proper distance.

Problem 11.4 One can define the proper distance $(dl)^2$ as

$$(d\mathbf{l})^2 \equiv (d\mathbf{s})^2 + (cdt)^2 \qquad ; \text{ proper distance}$$
$$\equiv (dl)^2$$

(a) Show this quantity is *invariant* under transformations between the global inertial laboratory frame I and the LF^3.

(b) Show that at a given instant in time, when the global inertial laboratory frame and the global LF^3 coincide, this becomes the ordinary spatial separation

$$(d\mathbf{l})^2 = (d\vec{l})^2 \qquad ; \text{ given instant}$$

(c) For a propagating light signal the interval vanishes $(d\mathbf{s})^2 = 0$. Show that the proper distance dl traveled by the light is related to the time interval dt by

$$dl = cdt \qquad ; \text{ light signal}$$

(d) Now imagine that the LF^3 has its origin at the observer and that at the instant a star emits a light signal, the star is a proper distance l away from the origin. Show the time Δt that the light signal takes to reach the origin is obtained from the integration of the result is (c) as

$$\Delta t = \frac{l}{c} \qquad ; \text{ time to go proper distance } l$$

Solution to Problem 11.4

(a) The time difference dt is identical in the global laboratory frame and LF^3. The interval $(ds)^2$ is identical in both frames. Hence the quantity

$$(d\,l)^2 \equiv (ds)^2 + (cdt)^2 \qquad ; \text{ proper distance}$$
$$\equiv (dl)^2$$

is *invariant* under transformations between the global inertial laboratory frame and the LF^3.[5]

(b) At a given instant in time, with $dt = 0$, the interval becomes the physical spatial length in the two frames

$$(d\,l)^2 = (d\vec{l}\,)^2 \qquad ; \text{ given instant}$$

(c) For a propagating light signal the interval vanishes $(ds)^2 = 0$. It then follows from the definition in part (a) that the proper distance dl traveled by the light is related to the time interval dt by

$$dl = cdt \qquad ; \text{ light signal}$$

(d) Imagine that the LF^3 has its origin at the observer and that at the instant a star emits a light signal, the star is a proper distance l away from the origin. The time Δt that the light signal takes to reach the origin can be obtained by integrating the result in (c) in the LF^3. The result is

$$\Delta t = \frac{l}{c} \qquad ; \text{ time to go proper distance } l$$

This is now identical to the time interval in the laboratory frame.

Problem 11.5 (a) Let t_e be the time of emission of the radiation and the present time $t_p = t_e + \Delta t$ that of observation. Show Eq. (11.68) is equivalent

---

[5]Note that in the LF^3, this definition of proper distance appropriately reduces to $(dl)^2 = (d\bar{l})^2$.

to the following expression for the redshift in the present cosmology

$$\frac{\lambda(t_p, t_e)}{\lambda(t_e)} = \left(\frac{t_p - t_0}{t_e - t_0}\right)^{2/3} = \left(\frac{t_p - t_0}{t_p - \Delta t - t_0}\right)^{2/3}$$

$$\Delta t = \frac{l}{c} = t_p - t_e$$

Here l is the physical spatial distance of the source at the time of emission.

(b) Hence conclude that there is an infinite redshift for light that comes from the horizon where $t_e \to t_0$ and $l \to D_H$.

Solution to Problem 11.5

(a) Equation (11.68) is re-written as

$$\frac{\lambda}{\lambda_0} = \frac{(t + \Delta t - t_0)^{2/3}}{(t - t_0)^{2/3}} \qquad ; \Delta t = \frac{l}{c}$$

where λ_0 is the wavelength at the time t.[6] Now identify the time of emission and the present time according to

$$t = t_e \qquad\qquad ; t_p = t_e + \Delta t$$

Hence

$$\frac{\lambda(t_p, t_e)}{\lambda(t_e)} = \left(\frac{t_p - t_0}{t_e - t_0}\right)^{2/3} = \left(\frac{t_p - t_0}{t_p - \Delta t - t_0}\right)^{2/3}$$

$$\Delta t = \frac{l}{c} = t_p - t_e$$

(b) Let the time of emission go back to the beginning at t_0, in which case Δt becomes the age of the universe, and l is the distance to the horizon

$$t_e \to t_0 \qquad ; \Delta t \to t_p - t_0 \qquad ; l \to c(t_p - t_0) = D_H$$

This is as far as we can see at the present time. One concludes that there is an infinite redshift for this ancient light

$$\frac{\lambda(t_p, t_e)}{\lambda(t_e)} \to \infty \qquad ; t_e \to t_0$$

---

[6] An unfortunate notation: Eq. (11.66) implies that in this expression $\lambda_0 = \lambda(t)$ where t is the time of emission; it is *not* $\lambda(t_0)$.

Problem 11.6 An object on the horizon emitted light that gets to us now, and when the light was emitted the object was a physical spatial distance D_H away. What is the physical distance of the object now?

(a) Let t_e be the time of emission, and t_p the present time. Make use of Eqs. (11.60) and (11.67) to show that the physical spatial distance of an object at a fixed coordinate point (q^1, q^2) satisfies

$$\frac{l(t_p)}{l(t_e)} = \left(\frac{t_p - t_0}{t_e - t_0}\right)^{2/3}$$

(b) Consider an object a small distance εc inside the horizon so that $l(t_e) = D_H - \varepsilon c$. If the light gets to us at the present time, then the light was emitted at a time $t_e = t_0 + \varepsilon$. Show from part (a) that these quantities are related by

$$\frac{l(t_p)}{D_H - \varepsilon c} = \left(\frac{t_p - t_0}{\varepsilon}\right)^{2/3}$$

or ; $\qquad l(t_p) \to D_H \left(\dfrac{t_p - t_0}{\varepsilon}\right)^{2/3} \qquad$; as $\varepsilon \to 0$

Hence conclude that an object on the horizon will be an *infinite distance away* at the time the light gets to us.[7]

Solution to Problem 11.6

(a) Equation (11.60) expresses the physical distance in terms of the expansion parameter in the metric

$$l = \Lambda(t) \left(\vec{q}^{\,2}\right)^{1/2} \qquad ; \text{ physical spatial distance}$$

The solution for that expansion parameter is given in Eq. (11.67)

$$\Lambda(t) = \gamma^{1/3}(t - t_0)^{2/3}$$

As in the previous problem, let t_e be the time of emission of the light and t_p be the present time. Then for a fixed coordinate separation in the laboratory frame, one has

$$\frac{l(t_p)}{l(t_e)} = \frac{(t_p - t_0)^{2/3}}{(t_e - t_0)^{2/3}}$$

---

[7]This correlates with the infinite redshift of the light from the horizon found in Prob. 11.5(b).

(b) Consider an object a small distance εc inside the horizon so that $l(t_e) = D_H - \varepsilon c$. If light emitted by the object gets to us at the present time t_p, then the light was emitted at a time $t_e = t_0 + \varepsilon$. From part (a), these quantities are related by

$$\frac{l(t_p)}{D_H - \varepsilon c} = \left(\frac{t_p - t_0}{\varepsilon}\right)^{2/3}$$

It follows that as $\varepsilon \to 0$

$$l(t_p) \to D_H \left(\frac{t_p - t_0}{\varepsilon}\right)^{2/3} \qquad ; \varepsilon \to 0$$

Hence we conclude that due to the expansion of the space, the object on the horizon will be an *infinite distance away* by the time the ancient light, traveling a time $\Delta t = t_p - t_0$ and distance $D_H = c(t_p - t_0)$, gets to us. With the aid of Fig. 11.9 in the text, one can now understand the corresponding infinite redshift of the light from the horizon found in Prob. 11.5(b).

Problem 11.7 The measured value of the Hubble constant in Eq. (11.83) corresponds to what value of the mass density ρ for the cosmology studied in this chapter?

Solution to Problem 11.7

The measured value of Hubble's constant quoted in Eq. (11.83) is

$$H_0 = 70 \pm 3 \ (\text{km/sec})/\text{Mpc}$$

In the present cosmology, Hubble's constant is related to the mass density through Eq. (11.36)

$$H_0 = h(t_p) = \frac{2}{3}[6\pi G \rho(t_p)]^{1/2}$$

One series of numerical values is given in the text in Eqs. (11.48)

$$\rho(t_p) = 2 \times 10^{-29} \ \text{gm/cm}^3$$
$$h(t_p) = 3.34 \times 10^{-18} \ \text{sec}^{-1} = 103 \ (\text{km/sec})/\text{Mpc}$$

To get the new value of $\rho(t_p)$, it is only necessary to scale the old one by the square of Hubble's constant

$$\rho(t_p)_{\text{new}} = 9.24 \times 10^{-30} \ \text{gm/cm}^3$$

If we use the proton mass from below Eq. (11.1)

$$m_p = 1.673 \times 10^{-24} \, \text{gm}$$

then this new mass density corresponds to a baryon density of

$$n_b(t_p) = 5.52 \, \text{protons/m}^3$$

Problem 11.8 There exists a cosmic microwave background (CMB) that fills all of space. It remains from the early hot era of the universe after the radiation decoupled and adiabatically expanded with time. The CMB exhibits an almost perfect *black-body spectrum* with an energy density

$$d\varepsilon_\gamma = \frac{8\pi\nu^2 \, d\nu}{c^3} \frac{h\nu}{e^{h\nu/k_B T} - 1}$$

and a current temperature $T(t_p) = 2.73\,°\text{K}$ [Ohanian (1995)].

(a) Show that when integrated over frequencies, this gives rise to a radiation density of

$$\varepsilon_\gamma = \frac{4\sigma}{c} T^4$$

Here σ is the Stefan-Boltzmann constant of Prob. 10.10, with the value $\sigma = 5.670 \times 10^{-5} \, \text{erg/sec-cm}^2\text{-}°\text{K}^4$.

(b) Define an effective radiation mass density by $\varepsilon_\gamma \equiv \rho_\gamma c^2$ and compute ρ_γ. Compare with the value of the matter (baryon) ρ appearing in Eqs. (11.48) and Prob. 11.7, and show that ρ_γ is a negligible source of gravity in the Einstein field equations.

Solution to Problem 11.8

(a) Integrate the given energy density

$$\varepsilon_\gamma = \int_0^\infty \frac{8\pi\nu^2 \, d\nu}{c^3} \frac{h\nu}{e^{h\nu/k_B T} - 1}$$

$$= \frac{(k_B T)^4}{\pi^2 (\hbar c)^3} \int_0^\infty \frac{x^3 \, dx}{e^x - 1} \qquad ; \, x \equiv \frac{h\nu}{k_B T}$$

Use the following definite integral[8]

$$\int_0^\infty \frac{x^3 \, dx}{e^x - 1} = \frac{\pi^4}{15}$$

---

[8] Any good symbolic manipulation program, for example Mathcad 7, now evaluates such definite integrals. Remember also that $\hbar = h/2\pi$.

Hence

$$\varepsilon_\gamma = \frac{\pi^2 (k_B T)^4}{15(\hbar c)^3}$$

The Stefan-Boltzmann constant is given in Prob. (10.10), and the above, by

$$\sigma = \frac{\pi^2 k_B^4}{60\hbar^3 c^2}$$
$$= 5.670 \times 10^{-5} \frac{\text{gm}}{\text{sec}^3 \text{-}^\circ\text{K}^4}$$

It follows that the CMB energy density is expressed in terms of the Stefan-Boltzmann constant by

$$\varepsilon_\gamma = \frac{4\sigma}{c} T^4$$

(b) Define the effective CMB mass density by

$$\rho_\gamma \equiv \frac{\varepsilon_\gamma}{c^2} = \frac{4\sigma}{c^3} T^4$$

Then with $T(t_p) = 2.73\,^\circ\text{K}$, and $c = 2.998 \times 10^{10}\,\text{cm/sec}$, one has

$$\rho_\gamma(t_p) = 4.67 \times 10^{-34} \frac{\text{gm}}{\text{cm}^3}$$

This is at least four orders of magnitude less than the baryon mass densities considered in Prob. 11.7.

Problem 11.9 One can define a *photon number density* spectrum dn_γ by dividing the energy density spectrum $d\varepsilon_\gamma$ of Prob. 11.8 by the energy per photon of $h\nu$. Thus

$$dn_\gamma \equiv \frac{d\varepsilon_\gamma}{h\nu} = \frac{8\pi\nu^2\,d\nu}{c^3} \frac{1}{e^{h\nu/k_B T} - 1}$$

(a) Integrate this expression and show the photon number density is then given by

$$n_\gamma = \left[\frac{2.404\, k_B^3}{\pi^2 (\hbar c)^3} \right] T^3$$

Note that this quantity goes as T^3.

The mass density in the present cosmology comes from baryons (protons) and can be written $\rho = m_b n_b$ where n_b is the baryon density. After "decoupling" in the era where the medium is so diffuse that photon reactions are no longer important, one can imagine that the mixture maintains a constant ratio of n_γ/n_b and merely adiabatically expands as the space stretches.

(b) Combine the result in part (a) with that in Prob. 11.1 to conclude that the time development of the temperature $T(t)$ and the scale factor in the metric $\Lambda(t)$ are thus related by

$$T^3(t)\Lambda^3(t) = \text{constant}$$

(c) Given the present value of $t_p - t_0$ in Eqs. (11.48), and $T(t_p)$ from Prob. 11.8, at what time in the past was k_BT at $1\,\text{MeV}$? (Note $k_B = 8.620 \times 10^{-5}\,\text{eV}/^\circ\text{K}$).[9] What was the corresponding mass density ρ at that time?

Solution to Problem 11.9

(a) A photon number density spectrum dn_γ can be defined by dividing the energy density spectrum $d\varepsilon_\gamma$ of Prob. 11.8 by the energy/photon of $h\nu$

$$dn_\gamma = \frac{8\pi\nu^2\,d\nu}{c^3}\frac{1}{e^{h\nu/k_BT}-1}$$

As in Prob. 11.8, integration of this expression gives

$$
\begin{aligned}
n_\gamma &= \int_0^\infty \frac{8\pi\nu^2\,d\nu}{c^3}\frac{1}{e^{h\nu/k_BT}-1}\\
&= \frac{(k_BT)^3}{\pi^2(\hbar c)^3}\int_0^\infty \frac{x^2\,dx}{e^x-1} \qquad ; x \equiv \frac{h\nu}{k_BT}
\end{aligned}
$$

Use the following[10]

$$\int_0^\infty \frac{x^2\,dx}{e^x-1} = 2\,\zeta(3) = 2\sum_{n=1}^\infty \frac{1}{n^3} = 2.404$$

Hence

$$n_\gamma = \left[\frac{2.404\,k_B^3}{\pi^2(\hbar c)^3}\right]T^3$$

---

[9]The "freezing out" of various reactions with falling temperature plays a central role in the study of element formation in the early universe (see, for example, [Kolb and Turner (1990)]).

[10]From Mathcad 7.

(b) It was shown in Prob. 11.1 that in the present cosmology

$$\rho(t) \, \Lambda^3(t) = \text{constant}$$

where the mass density comes from a given number of baryons $\rho = n_b m_b$. As stated in the problem, we assume that after "decoupling", the ratio of n_γ/n_b is maintained, where n_γ is the photon number density of part (a). The above relation can then be re-arranged to read

$$n_\gamma(t) \, \Lambda^3(t) = \text{constant}$$

Since $n_\gamma \propto T^3$, this leads to the conclusion that the temperature $T(t)$ and the scale factor in the metric $\Lambda(t)$ are related by[11]

$$T^3(t)\Lambda^3(t) = \text{constant}$$

(c) The solution in Eq. (11.39) gives $\Lambda(t) = \gamma^{1/3}(t - t_0)^{2/3}$. Therefore, from part (b), the temperature at the present time t_p, is related to that at the past time t_1 by

$$\frac{T(t_p)}{T(t_1)} = \left(\frac{t_1 - t_0}{t_p - t_0}\right)^{2/3}$$

Use the present time from Eq. (11.48),

$$t_p - t_0 = 2.00 \times 10^{17} \, \text{sec}$$

Also, use the temperatures

$$T_p = 2.73 \, ^\circ\text{K}$$

$$k_B T_1 = 1 \, \text{MeV} \qquad ; \; k_B = 8.620 \times 10^{-5} \, \frac{\text{eV}}{^\circ\text{K}}$$

This gives the following time interval $t_1 - t_0$ at which $k_B T_1 = 1 \, \text{MeV}$

$$t_1 - t_0 = 7.22 \times 10^2 \, \text{sec}$$

From Eq. (11.36), the mass density scales as $\rho(t) \propto (t - t_0)^{-2}$. Hence the coresponding mass density at the time t_1 is related to that at the present

---

[11]Note that it is because we know the relation of n_γ to the temperature that we are able to derive this relation.

time in Eqs. (11.48) by

$$\rho(t_p) = 2 \times 10^{-29} \, \frac{\text{gm}}{\text{cm}^3}$$

$$\rho(t_1) = \left(\frac{t_p - t_0}{t_1 - t_0}\right)^2 \rho(t_p) = 1.53 \, \frac{\text{gm}}{\text{cm}^3}$$

Chapter 12

Gravitational Radiation

Problem 12.1 (a) Expand the coordinate transformation matrix for the gravitational wave in Eq. (12.66) through first order in γ_{ij} [recall Eq. (12.62)], and then insert the time dependence of the metric in Eq. (12.67).

(b) Now explicitly carry out the transformation from the coordinates (x, y, z, ct) in the global inertial laboratory frame to the coordinates $(\bar{x}, \bar{y}, \bar{z}, c\bar{t})$ in the LF^3. Discuss.

Solution to Problem 12.1

(a) Equation (12.62) defines the angle χ in terms of the elements γ_{ij} [1]

$$(1 + \gamma_{xx})^{1/2}(1 + \gamma_{yy})^{1/2} \sin 2\chi \equiv \gamma_{xy}$$

An expansion through $O(\gamma)$ then gives

$$\chi = \frac{1}{2}\gamma_{xy} + O(\gamma^2)$$

The coordinate transformation matrix in Eq. (12.66) is

$$[\bar{a}]_{\nu\mu} = \bar{a}^{\nu}{}_{\mu}$$

$$[\bar{a}]_{\nu\mu} = \begin{bmatrix} (1 + \gamma_{xx})^{1/2} \cos\chi & (1 + \gamma_{yy})^{1/2} \sin\chi & 0 & 0 \\ (1 + \gamma_{xx})^{1/2} \sin\chi & (1 + \gamma_{yy})^{1/2} \cos\chi & 0 & 0 \\ 0 & 0 & 1 & 0 \\ 0 & 0 & 0 & 1 \end{bmatrix}$$

---

[1] Note $\gamma_{xy} = \gamma_{yx}$.

An expansion through $O(\gamma)$ then yields[2]

$$[\bar{a}]_{\nu\mu} = \begin{bmatrix} 1 + \gamma_{xx}/2 & \gamma_{xy}/2 & 0 & 0 \\ \gamma_{xy}/2 & 1 + \gamma_{yy}/2 & 0 & 0 \\ 0 & 0 & 1 & 0 \\ 0 & 0 & 0 & 1 \end{bmatrix} + O(\gamma^2)$$

The time dependence of the metric for a wave traveling in the z-direction is given in Eq. (12.67)

$$\gamma_{ij} = h_{ij} \cos k(z - ct) \qquad ; (i, j) = (x, y)$$

where h_{ij} are constants. This gives

$$[\bar{a}]_{\nu\mu} = \begin{bmatrix} 1 & 0 & 0 & 0 \\ 0 & 1 & 0 & 0 \\ 0 & 0 & 1 & 0 \\ 0 & 0 & 0 & 1 \end{bmatrix} + \frac{1}{2}\begin{bmatrix} h_{xx} & h_{xy} & 0 & 0 \\ h_{xy} & h_{yy} & 0 & 0 \\ 0 & 0 & 0 & 0 \\ 0 & 0 & 0 & 0 \end{bmatrix} \cos k(z - ct) + O(\gamma^2)$$

(b) The transformation from the coordinates $q^\nu = (x, y, z, ct)$ in the global inertial laboratory frame to the coordinates $\bar{q}^\nu = (\bar{x}, \bar{y}, \bar{z}, c\bar{t})$ in the LF^3 is then given by

$$\bar{q}^\nu = \bar{a}^\nu{}_\mu q^\mu$$

In detail, this gives through $O(\gamma)$

$$\bar{x} = x + \frac{x}{2} h_{xx} \cos k(z - ct) + \frac{y}{2} h_{xy} \cos k(z - ct)$$

$$\bar{y} = y + \frac{y}{2} h_{yy} \cos k(z - ct) + \frac{x}{2} h_{xy} \cos k(z - ct)$$

$$\bar{z} = z$$

$$\bar{t} = t$$

All the action takes place in the transverse plane, as indicated in Eq. (12.57) and the accompanying discussion; see, in particular, Fig. (12.6) in the text.

Problem 12.2 Consider the lagrangian $L_{\text{NRL}}(\dot{x}, \dot{y}, \dot{z}; x, y, z; t)$ of Eq. (12.73) where the perturbation of the metric is given in Eq. (12.67).

---

[2]Recall that for the plane-wave solution, from Eqs. (12.55),

$$\gamma_{xx} = -\gamma_{yy}, \qquad \gamma_{xy} = \gamma_{yx} \qquad ; \text{ all others vanish}$$

(a) What are the consequences of the fact that the coordinates (x, y) are cyclic?

(b) What are the consequences of the fact that this lagrangian has an *explicit* dependence on the time?

Solution to Problem 12.2

(a) The lagrangian for particle motion in a gravitational plane wave moving in the z-direction is given in the NRL by Eq. (12.73)

$$L_{\text{NRL}} = -mc^2 + \frac{m}{2} \left[\dot{z}^2 + (1 + \gamma_{xx})\dot{x}^2 + (1 + \gamma_{yy})\dot{y}^2 + 2\gamma_{xy}\,\dot{x}\dot{y} \right]$$

The metric perturbation, studied in Prob. 12.1, is given by

$$\gamma_{ij} = h_{ij} \cos k(z - ct) \qquad ; (i, j) = (x, y)$$

The coordinates $q^i = (x, y)$ are cyclic (they do not appear in L). It follows from Lagrange's equations that the corresponding canonical momenta are constants of the motions

$$p_i = \frac{\partial L}{\partial(\partial L/\partial \dot{q}^i)} \qquad ; q^i = (x, y)$$

$$\frac{dp_i}{dt} = 0$$

This is evident from the equations of motion derived from L in Eqs. (12.74)[3]

$$\frac{d}{dt}\left[(1 + \gamma_{xx})\dot{x} + \gamma_{xy}\,\dot{y} \right] = 0$$

$$\frac{d}{dt}\left[(1 + \gamma_{yy})\dot{y} + \gamma_{xy}\,\dot{x} \right] = 0$$

$$\left[\frac{d}{dt}\dot{z} + \frac{k}{2}\left(\gamma_{xx}\,\dot{x}^2 + \gamma_{yy}\,\dot{y}^2 + 2\gamma_{xy}\,\dot{x}\dot{y} \right) \tan k(z - ct) \right] = 0$$

(b) The lagrangian in part (a) has an *explicit* dependence on the time so that $\partial L/\partial t \neq 0$. As a consequence, the gravitational wave can pump energy to and from the particle, and the particle's energy is not conserved.

Problem 12.3 Write, or obtain, a program to solve numerically the three coupled non-linear differential Eqs. (12.74) for particle motion in the presence of a gravitational wave. Use dimensionless variables $(kx, ky, kz, \omega t)$ with $\omega = kc$, and assume small values for the constants (h_{xx}, h_{xy}). Investigate the subsequent motion for various initial conditions.

---

[3] Note the discussion in the text following these equations.

Solution to Problem 12.3

We assume a gravitational wave moving in the z-direction with a metric of the form in Eqs. (12.67) and Probs. 12.1–12.2. The motion of the particle in the transverse (x, y)-plane is then governed by the first two of Eqs. (12.74) and Prob. 12.2(a), which imply two constants of the motion (p_x, p_y). In matrix form, at $z = 0$, these equations become (we suppress the mass m)

$$\left[\begin{pmatrix} 1 & 0 \\ 0 & 1 \end{pmatrix} + \begin{pmatrix} h_{xx} & h_{xy} \\ h_{xy} & -h_{xx} \end{pmatrix} \cos \omega t \right] \begin{bmatrix} \dot{x} \\ \dot{y} \end{bmatrix} = \begin{bmatrix} p_x \\ p_y \end{bmatrix} \qquad ; z = 0$$

We assume (h_{xx}, h_{xy}) are small and work to first-order in these quantities. The above equations are then inverted according to

$$\begin{bmatrix} \dot{x} \\ \dot{y} \end{bmatrix} = \left[\begin{pmatrix} 1 & 0 \\ 0 & 1 \end{pmatrix} - \begin{pmatrix} h_{xx} & h_{xy} \\ h_{xy} & -h_{xx} \end{pmatrix} \cos \omega t \right] \begin{bmatrix} p_x \\ p_y \end{bmatrix} + O(h^2)$$

This relation is immediately integrated on the time to give

$$\begin{bmatrix} x \\ y \end{bmatrix} = \left[\begin{pmatrix} 1 & 0 \\ 0 & 1 \end{pmatrix} \frac{\omega t}{\omega} - \begin{pmatrix} h_{xx} & h_{xy} \\ h_{xy} & -h_{xx} \end{pmatrix} \frac{\sin \omega t}{\omega} \right] \begin{bmatrix} p_x \\ p_y \end{bmatrix} + \begin{bmatrix} x_0 \\ y_0 \end{bmatrix}$$

Mathcad 7 can now be used to compute an illustrative trajectory with $h_{xx} = h_{xy} = 0.05$ and initial values

$$\begin{bmatrix} x_0 \\ y_0 \end{bmatrix} = \begin{bmatrix} 1 \\ 0 \end{bmatrix} \qquad ; \begin{bmatrix} p_x/\omega \\ p_y/\omega \end{bmatrix} = \begin{bmatrix} 0 \\ 2 \end{bmatrix}$$

The result is shown in Fig. 12.1. This is the trajectory the record keeper sees on his or her screen. The particle starts off at $x = 1$ moving in the y-direction. In the absence of a gravitational wave, it simply moves up along the vertical straight line. In the presence of this gravitational wave, the particle oscillates back and forth along that line.[4] In principle, this motion could serve as a gravitational wave detector.

To first order in (h_{xx}, h_{xy}), the remaining equation for the z-motion of the particle becomes

$$\frac{d^2 z}{dt^2} + \frac{k}{2} \left(h_{xx} p_x^2 - h_{xx} p_y^2 + 2 h_{xy} p_x p_y \right) \sin k(z - ct) = 0$$

To this order, one can use the unperturbed motion to evaluate z in the argument of the sine. We were unable to find anything interesting for the z-motion with the above set of initial conditions in the transverse plane.[5]

---

[4] Compare Fig. 12.9 in the text.

[5] Readers are urged to look a little harder!

PARTICLE TRAJECTORY IN GRAVITATIONAL WAVE

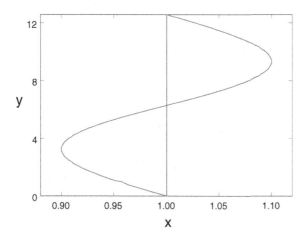

Fig. 12.1 Particle trajectory the record keeper sees in the transverse (x, y)-plane at $z = 0$ in a gravitational wave moving in the z-direction with $h_{xx} = h_{xy} = 0.05$. The particle starts at $(x_0, y_0) = (1, 0)$ with $(p_x/\omega, p_y/\omega) = (0, 2)$, and moves up in the positive-y direction. The vertical straight line is the trajectory in the absence of the wave.

Problem 12.4 The physical separation l of two particles a fixed coordinate distance d from each other on the x-axis, for a gravitational plane wave moving in the z-direction, is given by Eq. (12.78) as

$$l = d + \frac{d}{2} h_{xx} \cos k(z - ct)$$

(a) Use $\mu \ddot{l}$ to show that the equivalent gravitational interparticle *force* on the pair F_g is given by

$$F_g = -\mu\omega^2 \frac{d}{2} h_{xx} \cos k(z - ct)$$

where μ is the reduced mass.

(b) Suppose the two particles in (a) are connected by a spring of equilibrium length l_0 and spring constant $k = \mu\omega_0^2$. Assume $z = 0$ for simplicity, and set $d \approx l_0$ in the driving term. Write Newton's second law for the relative motion of the pair, and show that the general solution to that equation is given by

$$\zeta(t) = A \cos \omega_0 t + B \sin \omega_0 t + \frac{1}{2} \frac{l_0 h_{xx} \omega^2}{(\omega^2 - \omega_0^2)} \cos \omega t \qquad ; \zeta \equiv l - l_0$$

Here (A, B) are constants.

(c) What is it that prevents the resonant contribution from actually becoming infinite?

Solution to Problem 12.4

(a) The physical separation l of two particles lying on the x-axis as the gravitational plane wave, moving in the z-direction, passes by is given in Eq. (12.78)

$$l = d + \frac{d}{2} h_{xx} \cos k(z - ct)$$

The resulting acceleration in the relative coordinate can be associated with a relative gravitational force through Newton's second law

$$F_g = \mu \ddot{l}$$

where μ is the reduced mass. Two time derivatives then give

$$F_g = -\mu \omega^2 \frac{d}{2} h_{xx} \cos k(z - ct)$$

(b) Suppose the two particles in (a) are connected by a spring of equilibrium length l_0 and spring constant $k = \mu \omega_0^2$. Newton's second law for the relative motion of the pair then becomes, with $\zeta \equiv l - l_0$

$$\mu \ddot{\zeta}(t) = F_g - k\zeta \qquad ; \zeta \equiv l - l_0$$

Assume $z = 0$ for simplicity, and set $d \approx l_0$ in the driving term. The above equation is then re-written as

$$\ddot{\zeta}(t) + \omega_0^2 \zeta(t) \approx -\frac{l_0 \omega^2}{2} h_{xx} \cos \omega t$$

This simple second-order, inhomogeneous, differential equation has the genera solution

$$\zeta(t) = A \cos \omega_0 t + B \sin \omega_0 t + \frac{1}{2} \frac{l_0 \omega^2}{(\omega^2 - \omega_0^2)} h_{xx} \cos \omega t$$

where (A, B) are constants. The response is *resonant* when the angular frequency $\omega = kc$ of the gravitational wave is equal to the intrinsic angular frequency ω_0 of the spring.

(c) There will always be some damping of the spring's motion $\propto \dot{\zeta}(t)$, neglected here, which will give the resonant response a finite width.

Problem 12.5 (a) Consider the *source* of the gravitational radiation. Suppose one has a given, localized energy-momentum tensor $T_{\mu\nu}(\vec{x}, t)$. Show that if the source is weak enough, Einstein's linearized field equations for gravitational radiation can be written *everywhere* as

$$\Box \gamma_{\mu\nu}(\vec{x}, t) = -\frac{16\pi G}{c^4} S_{\mu\nu}(\vec{x}, t)$$

where the source tensor $S_{\mu\nu}$ is given by

$$S_{\mu\nu} = T_{\mu\nu} - \frac{1}{2} T g_{\mu\nu}^0 \qquad ; T = T^\lambda{}_\lambda$$

What does "weak enough" mean?

(b) Insert a Fourier transform in the time

$$f(\vec{x}, t) = \frac{1}{2\pi} \int_{-\infty}^{\infty} \tilde{f}(\vec{x}, \omega) e^{-i\omega t} \, d\omega$$

and show that the inhomogeneous wave equation in (a) reduces to the inhomogeneous Helmholtz equation

$$(\nabla^2 + k^2) \, \tilde{\gamma}_{\mu\nu}(\vec{x}, \omega) = -\frac{16\pi G}{c^4} \tilde{S}_{\mu\nu}(\vec{x}, \omega)$$

Here $k = \omega/c$.

(c) Show that the driven solution to this equation can be written with the aid of the Green's function for the Helmholtz equation as[6]

$$\tilde{\gamma}_{\mu\nu}(\vec{x}, \omega) = \frac{16\pi G}{c^4} \int \frac{e^{ik|\vec{x} - \vec{y}|}}{4\pi |\vec{x} - \vec{y}|} \tilde{S}_{\mu\nu}(\vec{y}, \omega) d^3y$$

(d) Explore the consequences of the solution in (c).

Solution to Problem 12.5

(a) The Einstein field equations can be written in the form of Eqs. (7.14)

$$R_{\mu\nu} = \kappa \left(T_{\mu\nu} - \frac{1}{2} T g_{\mu\nu} \right) \qquad ; T = T^\lambda{}_\lambda$$

Here, from Eq. (10.55),

$$\kappa = \frac{8\pi G}{c^4}$$

---

[6]See [Fetter and Walecka (2003)], pp. 314–317]. Recall that one takes Re at the end.

Consider a perturbation of the metric

$$g_{\mu\nu} = g^0_{\mu\nu} + \gamma_{\mu\nu}$$

Provided that the auxiliary condition in Eq. (12.35) is satisfied,[7]

$$\frac{\partial}{\partial x_\nu}\psi_\mu{}^\nu = 0$$

$$\psi_\mu{}^\nu = \gamma_\mu{}^\nu - \frac{1}{2}\gamma\,\delta_\mu{}^\nu \qquad\quad ; \gamma = \gamma^\lambda{}_\lambda$$

then the Ricci tensor can be written as in Eq. (12.36)

$$R_{\mu\nu} = -\frac{1}{2}\Box\gamma_{\mu\nu}$$

A combination of these relations gives the field equations as

$$\Box\gamma_{\mu\nu}(\vec{x},t) = -\frac{16\pi G}{c^4}S_{\mu\nu}(\vec{x},t)$$

$$S_{\mu\nu} = T_{\mu\nu} - \frac{1}{2}Tg^0_{\mu\nu}$$

Note that we are always working only through $O(\gamma_{\mu\nu})$, and for these last relations to be valid, the source must be *weak enough* so that corrections of $O(\gamma^2_{\mu\nu})$ are negligible, which is also why we can use $g^0_{\mu\nu}$ in the source. If this is true, then these field equations hold *everywhere* in a flat-space background.

(b) Now insert a Fourier transform in the time for both the metric and source

$$f(\vec{x},t) = \frac{1}{2\pi}\int_{-\infty}^{\infty}\tilde{f}(\vec{x},\omega)e^{-i\omega t}\,d\omega$$

The inhomogeneous wave equation in (a) then reduces to the inhomogeneous Helmholtz equation involving only the spatial coordinate $\vec{x}$

$$(\nabla^2 + k^2)\,\tilde{\gamma}_{\mu\nu}(\vec{x},\omega) = -\frac{16\pi G}{c^4}\tilde{S}_{\mu\nu}(\vec{x},\omega)$$

Here $k = \omega/c$.

(c) The Green's function for the inhomogeneous Helmholtz equation, familiar from scattering theory, is

$$G_k(\vec{x},\vec{y}) = \frac{e^{ik|\vec{x}-\vec{y}|}}{4\pi|\vec{x}-\vec{y}|}$$

---

[7]See the next Prob. 12.6.

Hence the solution to the scattering equation in (b) can be written in terms of the source as[8]

$$\tilde{\gamma}_{\mu\nu}(\vec{x},\omega) = \frac{16\pi G}{c^4} \int \frac{e^{ik|\vec{x}-\vec{y}|}}{4\pi|\vec{x}-\vec{y}|} \tilde{S}_{\mu\nu}(\vec{y},\omega) d^3y$$

(d) To compute any interesting physical quantities, we need an energy flux. Since everything follows from a lagrangian density (see the *Aside* at the end of this problem), we construct a useful *effective* field theory. The following Prob. 12.6 demonstrates that the subsidiary condition holds for all times, if it holds initially; we assume this to be the case. A contraction of the indices then reduces the field equations in part (a) to the form

$$\Box \gamma(\vec{x},t) = -2\kappa\, S(\vec{x},t) \qquad\qquad ; \gamma = \gamma_\mu{}^\mu, \qquad S = S_\mu{}^\mu$$

Now use the analogy to massless scalar meson field theory. A lagrangian density for which this is the field equation is, in flat Minkowski space,

$$\mathcal{L}_{\text{eff}} = -\frac{1}{4\kappa} g^{\mu\nu} \left(\frac{\partial\gamma}{\partial x^\mu}\right) \left(\frac{\partial\gamma}{\partial x^\nu}\right) + S\gamma$$

As justification, we note the following:

- Lagrange's equation yields the correct field equation

$$\Box\gamma = -2\kappa S$$

- γ is now appropriately dimensionless (see Prob. 12.7);
- The interaction term $S\gamma$ then has the correct dimensions;
- The energy flux $\vec{S}$ (see below) has the correct dimensions.
- The factor $1/2\kappa$ in front of the sourceless contribution is the same as in the Einstein-Hilbert lagrangian [see Eq. (13.3)].

Although the more rigorous and fundamental Einstein-Hilbert lagrangian density for the full gravitational field is derived in chapter 13,[9] we are now in a position to calculate *any* properties of the gravitational radiation in this effective linearized problem.

The final result in part (c) reads

$$\tilde{\gamma}(\vec{x},\omega) = 2\kappa \int \frac{e^{ik|\vec{x}-\vec{y}|}}{4\pi|\vec{x}-\vec{y}|} \tilde{S}(\vec{y},\omega)\, d^3y$$

---

[8]The sign in the statement of the problem in the text has been corrected here.
[9]And we may here have mutilated the polarizations (see the *Aside* in Prob. 12.6).

The asymptotic form of this expression for $|\vec{x}| = r \to \infty$, and $\vec{y}$ confined to the source, follows from elementary scattering theory[10]

$$\tilde{\gamma}(\vec{x}, \omega) \to f(k, \theta, \phi) \frac{e^{ikr}}{r} \qquad ; \quad |\vec{x}| = r \to \infty$$

$$f(k, \theta, \phi) = \frac{\kappa}{2\pi} \int e^{-i\vec{k}'\cdot\vec{y}} \, \tilde{S}(\vec{y}, \omega) \, d^3y$$

Here $\vec{k}'$ is a vector of magnitude $k = \omega/c$ that comes off at angles (θ, ϕ) relative to the z-axis.

The energy flux in a source-free region is, from the *Aside* below,

$$\frac{1}{c}\mathcal{S}_i = \mathcal{T}^{4i} = \frac{\partial \mathcal{L}}{\partial(\partial\gamma/\partial x^i)} \frac{\partial\gamma}{\partial(ct)}$$

This is computed to give[11]

$$\vec{\mathcal{S}} = -\frac{1}{2\kappa} \frac{\partial\gamma}{\partial t} \vec{\nabla}\gamma$$

Let us specialize to a monochromatic source and response, with a uniform time dependence $e^{-i\omega t}$. We are then dealing with the derivatives of the harmonic expression $\tilde{\gamma}\, e^{-i\omega t}$ that are complex. It is the *real part* of such quantities that represent physics. Furthermore, we are interested in the *time average* rate of energy emission. The calculation is then just that of Eqs. (45.43)–(45.45) in [Fetter and Walecka (2003)]. Write

$$\mathrm{Re}\,\alpha e^{-i\omega t} = \frac{1}{2}(\alpha e^{-i\omega t} + \alpha^\star e^{i\omega t})$$

$$\mathrm{Re}\,\beta e^{-i\omega t} = \frac{1}{2}(\beta e^{-i\omega t} + \beta^\star e^{i\omega t})$$

The time average, indicated as $\langle \cdots \rangle$, is then given by

$$\left\langle (\mathrm{Re}\,\alpha e^{-i\omega t})(\mathrm{Re}\,\beta e^{-i\omega t}) \right\rangle = \frac{1}{4} \left\langle \alpha\beta e^{-2i\omega t} + \alpha\beta^* + \alpha^\star\beta + \alpha^\star\beta^\star e^{2i\omega t} \right\rangle$$

Since the time average of the exponentials vanishes, this is

$$\left\langle (\mathrm{Re}\,\alpha e^{-i\omega t})(\mathrm{Re}\,\beta e^{-i\omega t}) \right\rangle = \frac{1}{4}(\alpha\beta^* + \alpha^\star\beta) = \frac{1}{2}\mathrm{Re}\,(\alpha\beta^\star)$$

---

[10] See also [Walecka (2013)].
[11] Compare Eq. (48.70) in [Fetter and Walecka (2003)].

Suppose we are now given the following expression that holds asymptotically for $|\vec{x}| = r \to \infty$

$$\gamma = f(k, \theta, \phi)\frac{e^{i(kr - \omega t)}}{r} \qquad ; r \to \infty$$

It follows that on a big sphere far away from the source

$$\frac{\partial \gamma}{\partial t} = -i\omega f(k, \theta, \phi)\frac{e^{i(kr - \omega t)}}{r}$$

$$\mathbf{e}_r \cdot \vec{\nabla}\gamma = = ikf(k, \theta, \phi)\frac{e^{i(kr - \omega t)}}{r}$$

The time-average energy flux radiated into a given solid angle is then obtained from the above as

$$\langle \mathbf{e}_r \cdot \mathcal{S}\rangle r^2 d\Omega = \frac{\omega^2}{4\kappa c}|f(k, \theta, \phi)|^2 d\Omega$$

The amplitude $f(k, \theta, \phi)$ was obtained previously as

$$f(k, \theta, \phi) = \frac{\kappa}{2\pi}\int e^{-i\vec{k}' \cdot \vec{y}}\,\tilde{S}(\vec{y}, \omega)\,d^3y$$

and we now have the sought-after expression for the radiated power.[12]

We can go further. If the wavelength is long compared to the size of the source, one can start expanding the exponential inside the last integral:

---

[12]If there is a *distribution* of frequencies, and we use a Fourier series with a long period $\mathcal{T}$ rather than a Fourier transform to facilitate the time averaging, then the asymptotic form of γ as $r \to \infty$ becomes

$$\gamma = \frac{1}{2r}\left(\frac{2\pi}{\mathcal{T}}\right)^{1/2}\sum_{n=1}^{\infty}\left[f(k_n, \theta, \phi)e^{i\omega_n(r-ct)/c} + f^*(k_n, \theta, \phi)e^{-i\omega_n(r-ct)/c}\right]; \ \omega_n = \frac{2\pi n}{\mathcal{T}}$$

This expression is real, contains outgoing waves, has both positive and negative frequencies, and neglects the $n = 0$ term, which is independent of time and cannot radiate. We assume a similar expansion applies to the source. A repetition of the argument for time averaging then leads to

$$\langle \mathbf{e}_r \cdot \mathcal{S}\rangle r^2 d\Omega = \frac{2\pi}{\mathcal{T}}\sum_{n=1}^{\infty}\left[\frac{\omega_n^2}{4\kappa c}|f(k_n, \theta, \phi)|^2\right]d\Omega$$

For large $\mathcal{T}$ the sum over n can be converted to an integral over ω in the usual fashion $(2\pi/\mathcal{T})\sum_n \to \int d\omega$, and with a continuum of frequencies the radiated power becomes

$$\langle \mathbf{e}_r \cdot \mathcal{S}\rangle r^2 d\Omega = \int_0^{\infty} d\omega \left[\frac{\omega^2}{4\kappa c}|f(k, \theta, \phi)|^2\right]d\Omega$$

(1) We first observe that if $P/c^2 \ll \rho$ for the source, which we assume to be the case, then the integral over all space of $S(\vec{y}, t)$, is just the total mass of the source, which is independent of time and cannot radiate[13]

$$\int S(\vec{y}, t)\, d^3 y = Mc^2 \qquad \text{; mass of source}$$

There is *no* monopole gravitational radiation. We can build this condition in explicitly by subtracting 1 from the exponential

$$f(k, \theta, \phi) = \frac{\kappa}{2\pi} \int \left(e^{-i\vec{k}' \cdot \vec{y}} - 1 \right) \tilde{S}(\vec{y}, \omega)\, d^3 y \quad \text{; no monopole radiation}$$

(2) The first term in the new expansion *also* vanishes due to the definition of the C-M of the source

$$\int \vec{y}\, S(\vec{y}\, t)\, d^3 y = 0 \qquad \text{; C-M of source}$$

There is also *no* dipole gravitational radiation.

(3) The first non-vanishing contribution to the emitted power therefore arises from the quadrupole term

$$f(k, \theta, \phi) = -\frac{\kappa \omega^2}{4\pi c^2} \int (\mathbf{e}_r \cdot \vec{y}) (\mathbf{e}_r \cdot \vec{y})\, \tilde{S}(\vec{y}, \omega)\, d^3 y + O[\kappa (k\mathcal{R})^3] \qquad \text{; } \mathbf{e}_r = \frac{\vec{k}'}{k}$$

Here the $\mathcal{R}$ in the neglected term characterizes the size of the emitter.

We are now in a position to calculate the power radiated from an arbitrary localized source! It remains only for readers to model $\tilde{S}(\vec{y}, \omega)$ and calculate $f(k, \theta, \phi)$.

****

(*Aside*) We review some results from continuum mechanics in flat Minkowski space, where $q(\vec{x}, t)$ is a Lorentz scalar field.[14] Given the lagrangian density $\mathcal{L}(q, \partial q/\partial x^\mu; x^\mu)$, Lagrange's equation then reads

$$\frac{\partial}{\partial x^\mu} \frac{\partial \mathcal{L}}{\partial(\partial q/\partial x^\mu)} = \frac{\partial \mathcal{L}}{\partial q} \qquad \text{; Lagrange's eqn}$$

---

[13] From Eq. (6.108), if $P/c^2 \ll \rho$, then $S = -T_\lambda{}^\lambda = \rho c^2$. More generally, an oscillating monopole cannot emit a helicity-2 graviton (see footnote 20 on p. 161).

[14] See Probs. 13.7 and 13.11.

If $\mathcal{L}$ has no explicit dependence on x^μ so that $\partial\mathcal{L}/\partial x^\mu = 0$, then there is a conserved energy-momentum tensor

$$\frac{\partial T^{\mu\nu}}{\partial x^\nu} = 0 \qquad ; \text{ if } \frac{\partial\mathcal{L}}{\partial x^\mu} = 0$$

given by

$$T^{\mu\nu} = \mathcal{L}g^{\mu\nu} - \frac{\partial\mathcal{L}}{\partial(\partial q/\partial x^\nu)}\frac{\partial q}{\partial x_\mu} \qquad ; \text{ energy-momentum tensor}$$

The energy (hamiltonian) density $\mathcal{H}$ and energy flux $\vec{S}$ are obtained from the energy-momentum tensor as

$$\left(\frac{1}{c}\vec{S}, \mathcal{H}\right) = T^{4\nu}$$

They satisfy the continuity equation

$$\frac{\partial\mathcal{H}}{\partial t} + \vec{\nabla}\cdot\vec{S} = 0 \qquad ; \text{ energy conservation}$$

We verify the above expression for the hamiltonian density $\mathcal{H}$. The canonical momentum density is

$$\Pi_q = \frac{\partial\mathcal{L}}{\partial(\partial q/\partial t)}$$

It follows with the aid of the Lorentz metric in Eqs. (6.5) and (6.7) that

$$T^{44} = -\mathcal{L} + \Pi_q\frac{\partial q}{\partial t}$$

This is indeed the correct expression for $\mathcal{H}$

$$T^{44} = \mathcal{H}$$

****

Problem 12.6 (a) Start from the expressions in Prob. 12.5(a), and show that the quantity $\psi_\mu{}^\nu$ defined in Eq. (12.35) satisfies the following equation

$$\Box\psi_\mu{}^\nu = -\frac{16\pi G}{c^4}T_\mu{}^\nu$$

(b) Use the conservation of the energy-momentum tensor in Minkowski space to show that

$$\Box\left(\frac{\partial}{\partial x^\nu}\psi_\mu{}^\nu\right) = 0$$

Hence conclude that if the auxiliary condition is satisfied initially, it will continue to be satisfied.[15]

Solution to Problem 12.6

(a) The result in Prob. 12.5(a) for the perturbation of the metric with a gravitational wave is

$$\Box \gamma_{\mu\nu}(\vec{x}, t) = -\frac{16\pi G}{c^4}\left[T_{\mu\nu} - \frac{1}{2}T g_{\mu\nu}^0\right] \qquad ; T = T^\lambda{}_\lambda$$

We work through $O(\gamma)$, so indices can be raised and lowered with the Lorentz metric $g_{\mu\nu}^0$, which is independent of space-time.

The auxiliary condition is given in Eq. (12.35)

$$\frac{\partial}{\partial x_\nu}\psi_\mu{}^\nu = 0 \qquad ; \mu = 1, 2, 3, 4$$

$$\psi_\mu{}^\nu = \gamma_\mu{}^\nu - \frac{1}{2}\gamma\,\delta_\mu{}^\nu \qquad ; \gamma = \gamma^\lambda{}_\lambda$$

Raise the ν index on the wave equation

$$\Box \gamma_\mu{}^\nu = -\frac{16\pi G}{c^4}\left[T_\mu{}^\nu - \frac{1}{2}T\delta_\mu{}^\nu\right]$$

Now contract the indices[16]

$$\Box \gamma = -\frac{16\pi G}{c^4}[-T]$$

Hence

$$\Box \psi_\mu{}^\nu = -\frac{16\pi G}{c^4}T_\mu{}^\nu$$

(b) The energy-momentum tensor is conserved. Thus, after lowering the first index, Eq. (6.130) gives

$$\frac{\partial}{\partial x^\nu}T_\mu{}^\nu = 0 \qquad ; \text{energy-momentum conservation}$$

$$\mu = 1, \cdots 4$$

---

[15]This is the direct analog of the demonstration in E&M, through the use of Maxwell's equation $\Box A^\nu = -j^\nu$ and current conservation, that if the condition for the Lorentz gauge $\partial A^\nu / \partial x^\nu = 0$ is satisfied originally, it will continue to be satisfied.

[16]Note $T = T^\lambda{}_\lambda = T_\lambda{}^\lambda$, and $\gamma^\lambda{}_\lambda = \gamma_\lambda{}^\lambda$.

It follows that

$$\Box \left(\frac{\partial}{\partial x^\nu} \psi_\mu{}^\nu \right) = 0$$

We conclude that if the auxiliary condition is satisfied initially, it will continue to be satisfied.

****

(*Aside*) We can actually construct a driven solution to the wave equation for $\psi_\mu{}^\nu$. Introduce the four-dimensional Fourier transform of the source

$$T_\mu{}^\nu(x) = \int \frac{d^4 k}{(2\pi)^4} e^{ik \cdot x} \tilde{T}_\mu{}^\nu(k)$$

where

$$k \cdot x = \vec{k} \cdot \vec{x} - k_0 ct \qquad ; \ d^4 k = d^3 k \, dk_0$$

In terms of the Fourier transform, the statement of the conservation of the energy-momentum tensor becomes

$$\frac{\partial}{\partial x^\nu} T_\mu{}^\nu(x) = \int \frac{d^4 k}{(2\pi)^4} e^{ik \cdot x} ik_\nu \tilde{T}_\mu{}^\nu(k) = 0$$

Inversion of the Fourier transform leads to the conservation statement in momentum space

$$k_\nu \tilde{T}_\mu{}^\nu(k) = 0$$

A driven solution to the wave equation for $\psi_\mu{}^\nu$ is then given by

$$\psi_\mu{}^\nu(x) = \frac{16\pi G}{c^4} \int_C \frac{d^4 k}{(2\pi)^4} \frac{e^{ik \cdot x}}{k^2} \tilde{T}_\mu{}^\nu(k)$$

where the four-dimensional k^2 in the denominator is given by

$$k^2 = \vec{k}^2 - k_0^2$$

Here different contours C around the singularities define Green's functions with different boundary conditions.[17] It follows that

$$\frac{\partial}{\partial x^\nu} \psi_\mu{}^\nu(x) = \frac{16\pi G}{c^4} \int_C \frac{d^4 k}{(2\pi)^4} \frac{e^{ik \cdot x}}{k^2} ik_\nu \tilde{T}_\mu{}^\nu(k)$$

$$= 0$$

---

[17]See, for example, [Walecka (2010)]; the above is re-written as $\psi_\mu{}^\nu(x) = (16\pi G/c^4)$ $\int d^4 x' \mathcal{G}(x - x') T_\mu{}^\nu(x')$ with $\mathcal{G}(x - x') = \int_C e^{ik \cdot (x - x')} d^4 k/(2\pi)^4 k^2$.

With this solution, the four-divergence of $\psi_\mu{}^\nu(x)$ vanishes everywhere for all times.

We can now use this analysis to try to deal with the polarization of the gravitational radiation. Since the auxiliary condition is satisfied everywhere, we have the inhomogeneous wave equation

$$\Box \gamma_{\mu\nu} = -\frac{16\pi G}{c^4} S_{\mu\nu}(x)$$

$$S_{\mu\nu} = T_{\mu\nu} - \frac{1}{2}T g^0_{\mu\nu}$$

Written in terms of a spatial dyadic, the solution to this wave equation is given as above

$$\underline{\underline{\gamma}}(x) = \frac{16\pi G}{c^4} \int_C \frac{d^4k}{(2\pi)^4} \frac{e^{ik\cdot x}}{k^2} \underline{\underline{\tilde{S}}}(k)$$

We know from the calculation of the rate for photon emission that for a given $\mathbf{k}$, it is the spherical basis vectors that project the unit helicity of the photon from the vector source density

$$\mathbf{e}_{\mathbf{k},\pm 1} = \mp \frac{1}{\sqrt{2}}(\mathbf{e}_{\mathbf{k},1} \pm i\,\mathbf{e}_{\mathbf{k},2})$$

where $(\mathbf{e}_{\mathbf{k},1}, \mathbf{e}_{\mathbf{k},2}, \mathbf{k}/|\mathbf{k}|)$ form an orthonormal triad.[18]

We cheat a little and assume we can compute the rate for the emission of a helicity-2 graviton from an analogous expression

$$\gamma_{\pm 2}(x) = \frac{16\pi G}{c^4} \int_C \frac{d^4k}{(2\pi)^4} \frac{e^{ik\cdot x}}{k^2} \tilde{S}_{\pm 2}(k)$$

where the source is projected according to[19]

$$\tilde{S}_{\pm 2}(k) = \mathbf{e}_{\mathbf{k},\pm 1} \cdot \underline{\underline{\tilde{S}}} \cdot \mathbf{e}_{\mathbf{k},\pm 1} = \frac{1}{2}\left[\tilde{S}_{11} - \tilde{S}_{22} \pm i(\tilde{S}_{12} + \tilde{S}_{21}) \right]$$

It then follows that to deal with the polarization of the gravitational radiation, one simply has to replace

$$|f|^2 \to |f_{+2}|^2 + |f_{-2}|^2$$

in the calculation of the time-average energy flux radiated into a given solid angle in Prob. 12.5.

****

---

[18]See appendix A in [Walecka (2010)].

[19]We could include a Clebsch-Gordan coefficient; however, $\langle 1, \pm 1, 1, \pm 1 | 112, \pm 2 \rangle = 1$.

The previous Probs. 12.5–12.6 describe gravitational radiation from a source in the linearized theory. The calculation of gravitational radiation in the full non-linear theory of general relativity goes beyond the scope of the text. The next problem, though, does present some general considerations.

Problem 12.7 In the next chapter we will relate the source to the motion of massive bodies. Here we simply make some dimensional arguments on the source. The power $\mathcal{P}$ is the total energy radiated per unit time. The dimensions of power are $[ml^2t^{-3}]$. The dimensions of Newton's constant G are $[m^{-1}l^3t^{-2}]$. The angular frequency of the source ω has dimensions of $[t^{-1}]$, and the ratio ω/c provides an inverse length $[l^{-1}]$.

(a) The power radiated must involve a factor of G, which enters into Einstein's field equations and governs the strength of coupling of gravity to mass.

(b) There is no *dipole* radiation since there will never be any net motion of the mass of an isolated system relative to its *center* of mass. Thus the leading multipole for radiation whose wavelength is long compared to the size of the system will be *quadrupole*.[20] The quadrupole moment depends on a mean value of the square of the radial size of the system, and thus the radiated power will also depend on the square of a (now dimensionless) radiating quadrupole moment

$$\left(\ddot{Q}\right)^2 \propto \left[\left(\frac{\omega}{c}\right)^2 \langle r^2 \rangle\right]^2$$

Thus the total power radiated goes as

$$P_{\text{rad}} \propto GM^2c\left(\frac{\omega}{c}\right)^6 \left(\langle r^2 \rangle\right)^2$$

(c) The shrinking of the orbit with time of the Hulse-Taylor pulsar, evidently a binary neutron star, is a beautiful example of a star system *emitting* gravitational radiation [Wiki (2006)]. Use the following relevant values, as well as those of Eqs. (10.87), to obtain a number for the above estimate for P_{rad}

$$M \approx 2M_\odot \qquad ; \langle r^2 \rangle \approx (2R_\odot)^2 \qquad ; \tau = 2\pi/\omega \approx 8\,\text{hrs}$$

---

[20] Although we will not demonstrate it here, the quantum of gravitational radiation, the *graviton*, is a spin-2 particle corresponding to the tensor nature of the gravitational field, and since it is massless, it can have only the maximum and minimum values of the helicity $\lambda = \pm 2$ (just as a photon, the massless spin-1 quantum of the vector electromagnetic field, has helicity $\lambda = \pm 1$). There is no monopole gravitational radiation in this theory. The factors in the estimate view graviton emission in analogy with photon emission.

Compare with the measured value for the Hulse-Taylor pulsar of $P_{\text{rad}} \approx 10^{33}$ erg/sec.

(d) Verify that indeed $2R_\odot \ll \lambda$ in part (c).[21]

The direct *detection* of gravitational radiation is the goal of LIGO.

Solution to Problem 12.7

(a) As stated in the problem, the dimensions of power are $[ml^2t^{-3}]$. The dimensions of Newton's constant G are $[m^{-1}l^3t^{-2}]$. The angular frequency of the source ω has dimensions of $[t^{-1}]$, and the ratio ω/c provides an inverse length $[l^{-1}]$. Then

$$[ml^2t^{-3}] = [m^{-1}l^3t^{-2}][m^2][t^{-1}][l^{-1}]$$

Hence to obtain the correct dimensions, one must then have

$$\mathcal{P} \propto GM^2 \frac{\omega^2}{c}$$

where M is a characteristic moving mass in the source.

(b) By definition, there will never be any dipole moment of the mass of an isolated system relative to its *center* of mass. Hence, there is no *dipole* gravitational radiation.[22] Thus the leading multipole for radiation whose wavelength is long compared to the size of the system will be *quadrupole*. The quadrupole moment depends on a mean value of the square of the radial size of the system, and thus the radiated power will also depend on the square of a (now dimensionless) radiating quadrupole moment

$$\left(\ddot{Q}\right)^2 \propto \left[\left(\frac{\omega}{c}\right)^2 \langle r^2\rangle\right]^2$$

Thus the total power radiated goes as

$$P_{\text{rad}} \propto GM^2c\left(\frac{\omega}{c}\right)^6 \langle r^2\rangle^2$$

(c) As stated in the problem, the shrinking of the orbit with time of the Hulse-Taylor pulsar, evidently a binary neutron star, is a beautiful example of a star system *emitting* gravitational radiation. We use the given values

$$M \approx 2M_\odot \qquad ; \ \langle r^2\rangle \approx (2R_\odot)^2 \qquad ; \ \tau = 2\pi/\omega \approx 8\,\text{hrs}$$

---

[21] It is also interesting to compare with Prob. 7.5(a).

[22] In contrast to electromagnetic radiation, where the most common situation is to have a dipole moment of the charge relative to the center-of-mass.

as well as those of Eqs. (10.87)

$$c = 3.00 \times 10^{10} \, \text{cm/sec}$$
$$G = 6.67 \times 10^{-8} \, \text{cm}^3/\text{gm-sec}^2$$
$$M_\odot = 1.99 \times 10^{33} \, \text{gm} \qquad ; R_\odot = 6.96 \times 10^{10} \, \text{cm}$$

This gives the following estimate for P_{rad}

$$P_{\text{rad}} \approx 1.76 \times 10^{30} \, \text{ergs/sec} \qquad ; \text{estimate}$$

This simple dimensional estimate compares favorably with the observed value for the Hulse-Taylor pulsar[23]

$$P_{\text{rad}} \approx 10^{33} \, \text{erg/sec} \qquad ; \text{measured}$$

(d) We compute $kR = 2\pi R/\lambda = \omega R/c$

$$kR = \frac{\omega}{c} 2R_\odot$$
$$= 1.01 \times 10^{-3} \ll 1$$

Indeed $2R_\odot \ll \lambda$, and the wavelength is long compared to the size of the system in part (c).

It is also interesting to compare the estimated parameters of the orbit of the binary neutron star with the result for circular motion in the Schwarzschild metric, as well as in the newtonian limit, in Prob. 7.5(a)

$$\omega^2 = \frac{MG}{a^3}$$

From above, for the binary neutron star

$$\omega = 2.18 \times 10^{-4} \, \text{sec}^{-1}$$
$$\left[\frac{(2M_\odot)G}{(2R_\odot)^3} \right]^{1/2} = 3.14 \times 10^{-4} \, \text{sec}^{-1}$$

These are pretty close.[24]

---

[23]It is amusing to note that, although unobtainable from dimensional analysis, Prob. 12.5 implies the effective coupling is really $16\pi G$. The estimate is then $P_{\text{rad}} \approx 8.85 \times 10^{31}$ ergs/sec, which is now within one order of magnitude of the measured value.

[24]If the binary star is two neutron stars, each of mass $M_\odot$, separated by $2R_\odot$, then in the newtonian limit, in the C-M system, the relative coordinate satisfies Newton's law

$$\mu \frac{d^2\vec{r}}{dt^2} = -M_\odot^2 G \frac{\vec{r}}{r^3} \qquad ; \mu = \frac{1}{2} M_\odot$$

where μ is the reduced mass. This yields the given expression for ω.

It is reported that LIGO has now directly *detected* gravitational waves! See, for example, [Sky and Telescope (2016)]. LIGO uses a long-baseline Michelson laser interferometer to detect the change in length of the arms induced by a passing gravitational wave [LIGO (2006); Sandberg (2016)].[25] Consider, for example, the situation in section 12.3, where one arm of the interferometer of physical length $d\bar{l}_x$ lies in the x-direction. Then from Eqs. (12.77)–(12.78), the physical length of the arm at any instant is

$$d\bar{l}_x = [(ds)^2]^{1/2} = \left[1 + \frac{h_{xx}}{2}\cos k(z - ct)\right] dx$$

The time it takes to go a distance $d\bar{l}_x$ in the LF^3 is obtained from $(ds)^2 = (d\bar{l}_x)^2 - (c\,d\bar{t}_x)^2 = 0$, and since the time part of the metric is unmodified, $d\bar{t}_x = dt_x$.[26] Hence

$$c\,dt_x = d\bar{l}_x$$

If the other arm of the spectrometer lies in the y-direction, then for this gravitational wave with $h_{yy} = -h_{xx}$

$$d\bar{l}_y = \left[1 - \frac{h_{xx}}{2}\cos k(z - ct)\right] dy$$
$$= c\,dt_y$$

The difference in optical pathlength, now calculated as in freshman physics, is exhibited in the interference pattern.

---

[25] There are actually two of them, one in the state of Washington and one in Louisiana.
[26] See the discussion of Eqs. (12.42).

Chapter 13

Special Topics

Problem 13.1 Obtain a symbolic manipulation program, and verify the results in Table B.1 for the affine connection in the case of the Robertson-Walker metric with $k \neq 0$. Start from the elements of the metric in Eqs. (B.2).

Solution to Problem 13.1

Introduce spatial spherical coordinates, so that

$$q^\mu = (q^1, q^2, q^3, q^4) = (r, \theta, \phi, ct)$$

The Robertson-Walker metric with $k \neq 0$ then generalizes the metric with $k = 0$ to the following, and its inverse, [Robertson (1935); Walker (1936)]

$$g_{\mu\nu} = \begin{bmatrix} \Lambda^2(t)/(1 - kr^2) & 0 & 0 & 0 \\ 0 & \Lambda^2(t)r^2 & 0 & 0 \\ 0 & 0 & \Lambda^2(t)r^2 \sin^2\theta & 0 \\ 0 & 0 & 0 & -1 \end{bmatrix}$$

$$g^{\mu\nu} = \begin{bmatrix} (1 - kr^2)/\Lambda^2(t) & 0 & 0 & 0 \\ 0 & 1/[\Lambda^2(t)r^2] & 0 & 0 \\ 0 & 0 & 1/[\Lambda^2(t)r^2 \sin^2\theta] & 0 \\ 0 & 0 & 0 & -1 \end{bmatrix}$$

The procedure is clear. The affine connection must first be computed from the metric

$$\Gamma^\lambda_{\mu\nu} = \frac{1}{2} g^{\lambda\sigma} \left[\frac{\partial g_{\sigma\nu}}{\partial q^\mu} + \frac{\partial g_{\sigma\mu}}{\partial q^\nu} - \frac{\partial g_{\mu\nu}}{\partial q^\sigma} \right]$$

The Ricci tensor must then be obtained from the affine connection (see Prob. 13.2).

Symbolic manipulation allows one to do algebra with symbols, which greatly simplifies long, repetitive, algebraic calculations. Here we use the Mathcad 11 symbolic manipulation program.[1] In the present case, the calculation is simplified by the fact that the metric and its inverse are *diagonal.* One first defines the functions $g_{\mu\nu}(\Lambda, k, r, \theta)$ and $g^{\mu\nu}(\Lambda, k, r, \theta)$ for $(\mu, \nu) = (1, 1), (2, 2), (3, 3), (4, 4)$. Since Mathcad does not permit the definition of functions with subscripts and superscripts, we call these $g\mu\nu$ and $h\mu\nu$, respectively. For example,

$$g33(\Lambda, k, r, \theta) := \Lambda^2 r^2 \sin^2 \theta \qquad ; \quad h33(\Lambda, k, r, \theta) := 1/[\Lambda^2 r^2 \sin^2 \theta]$$

The affine connection is then introduced. For example, $\Gamma^3_{23}(\Lambda, \dot{\Lambda}, k, r, \theta)$, which we call $\Gamma323$, is

$$\Gamma323(\Lambda, \dot{\Lambda}, k, r, \theta) := \frac{1}{2} h33(\Lambda, k, r, \theta) \left[\frac{\partial}{\partial \theta} g33(\Lambda, k, r, \theta) \right]$$

Here the derivatives are so simple they are inserted in the above by hand. For example,

$$\frac{\partial}{\partial \theta} g33(\Lambda, k, r, \theta) = 2\Lambda^2 r^2 \sin \theta \cos \theta$$

The affine connection is then evaluated symbolically in Mathcad, using the modifier "simplify". For, example

$$\Gamma323(\Lambda, \dot{\Lambda}, k, r, \theta) \text{ simplify} \to \cot \theta$$

This reproduces the result for $\Gamma^\phi_{\theta\phi}$ Table 13.1. A change of indices and derivatives then reproduces all the other entries.

Table 13.1   SUMMARY —Affine connection for Robertson-Walker metric with $k \neq 0$ and coordinates $(q^1, \cdots, q^4) = (r, \theta, \phi, ct)$. It is symmetric in its lower two indices so that $\Gamma^\lambda_{\mu\nu} = \Gamma^\lambda_{\nu\mu}$. In this table $\dot{\Lambda} = d\Lambda(t)/dt$. This is Table B.1 in the text.

$\Gamma^r_{rr} = kr/(1 - kr^2)$	$\Gamma^r_{\theta\theta} = -r(1 - kr^2)$	$\Gamma^r_{\phi\phi} = -r(1 - kr^2) \sin^2 \theta$
$\Gamma^r_{r4} = \dot{\Lambda}/\Lambda c$	$\Gamma^\theta_{\phi\phi} = -\sin \theta \cos \theta$	$\Gamma^\theta_{r\theta} = 1/r$
$\Gamma^\theta_{\theta4} = \dot{\Lambda}/\Lambda c$	$\Gamma^\phi_{\phi4} = \dot{\Lambda}/\Lambda c$	$\Gamma^\phi_{\theta\theta} = \cos \theta / \sin \theta$
$\Gamma^\phi_{r\phi} = 1/r$	$\Gamma^4_{rr} = \Lambda \dot{\Lambda}/(1 - kr^2)c$	$\Gamma^4_{\theta\theta} = r^2 \Lambda \dot{\Lambda}/c$
$\Gamma^4_{\phi\phi} = r^2 \Lambda \dot{\Lambda} \sin^2 \theta/c$	All others vanish	

---

[1] Readers can easily locate their own equivalent versions.

Problem 13.2 Obtain a symbolic manipulation program, and verify the results in Eqs. (B.6) for the Ricci tensor in the case of the Robertson-Walker metric with $k \neq 0$. Start from the elements of the affine connection in Table B.1.

Solution to Problem 13.2

The Ricci tensor is obtained from the affine connection as in Eq. (B.4)

$$R_{\mu\nu} = \frac{\partial}{\partial q^\lambda} \Gamma^\lambda_{\mu\nu} + \Gamma^\lambda_{\lambda\sigma}\Gamma^\sigma_{\mu\nu} - \left[\frac{\partial}{\partial q^\nu}\Gamma^\lambda_{\mu\lambda} + \Gamma^\lambda_{\nu\sigma}\Gamma^\sigma_{\mu\lambda}\right]$$

The Ricci tensor here was first calculated by hand, using the look-up Table 13.1. It was verified that the results given in Eqs. (B.5) are correct

$$R_{ij} = \frac{1}{c^2}\left[2\left(\frac{\dot\Lambda}{\Lambda}\right)^2 + \left(\frac{\ddot\Lambda}{\Lambda}\right) + \frac{2kc^2}{\Lambda^2}\right]g_{ij} \qquad ; (i,j) = (r,\theta,\phi)$$

$$R_{44} = -\frac{3}{c^2}\left(\frac{\ddot\Lambda}{\Lambda}\right)$$

$$R_{i4} = 0$$

Here $\dot\Lambda = d\Lambda/dt$ and $\ddot\Lambda = d^2\Lambda/dt^2$. Note the presence of g_{ij} on the r.h.s. of the first equation.

These results were then reproduced symbolically using Mathcad 11. One first enters the statement ORIGIN:= 1, ensuring that matrix indices start with 1, instead of the default value 0. The goal was then to exhibit all the indices on the affine connection as *subscripts*, so that they can be summed over symbolically. This was accomplished by first defining the affine connections as four 4×4 matrix functions. For example, Γ^r is given by

$\underline{\Gamma 1}(\Lambda, k, r, \theta, t) :=$

$$\begin{bmatrix} kr/(1-kr^2) & 0 & 0 & [1/\Lambda(t)]d\Lambda(t)/dt \\ 0 & -r(1-kr^2) & 0 & 0 \\ 0 & 0 & -r(1-kr^2)\sin^2\theta & 0 \\ [1/\Lambda(t)]d\Lambda(t)/dt & 0 & 0 & 0 \end{bmatrix}$$

Here dt stands for $d(ct)$. The four 4×4 matrix functions were then arranged into a 4×1 matrix of matrices

$$\underline{\Gamma}(\Lambda, k, r, \theta, t) := \begin{bmatrix} \underline{\Gamma 1}(\Lambda, k, r, \theta, t) \\ \underline{\Gamma 2}(\Lambda, k, r, \theta, t) \\ \underline{\Gamma 3}(\Lambda, k, r, \theta, t) \\ \underline{\Gamma 4}(\Lambda, k, r, \theta, t) \end{bmatrix} \qquad ; \text{matrix of matrices}$$

The required affine connection is now obtained symbolically through the appropriate selection of indices. for example, $\Gamma^p_{m,n}$ is given by

$$\{[\underline{\Gamma}(\Lambda, k, r, \theta, t)]_{p,1}\}_{m,n} \to \text{ gives } \Gamma^p_{m,n}(\Lambda, k, r, \theta, t)$$

The Ricci tensor consists of three contributions. The required sums for R_{mn} are

$$Amn(\Lambda, k, r, \theta, t) := \sum_{p=1}^{4} \sum_{q=1}^{4} [\{[\underline{\Gamma}(\Lambda, k, r, \theta, t)]_{p,1}\}_{p,q}\{[\underline{\Gamma}(\Lambda, k, r, \theta, t)]_{q,1}\}_{m,n} -$$
$$\{[\underline{\Gamma}(\Lambda, k, r, \theta, t)]_{p,1}\}_{n,q}\{[\underline{\Gamma}(\Lambda, k, r, \theta, t)]_{q,1}\}_{m,p}]$$

There are two derivative terms, which we write out. First

$$Bmn(\Lambda, k, r, \theta, t) := \frac{d}{dr}\{[\underline{\Gamma}(\Lambda, k, r, \theta, t)]_{1,1}\}_{m,n} +$$
$$\frac{d}{d\theta}\{[\underline{\Gamma}(\Lambda, k, r, \theta, t)]_{2,1}\}_{m,n} + \frac{d}{dt}\{[\underline{\Gamma}(\Lambda, k, r, \theta, t)]_{4,1}\}_{m,n}$$

Since there is no ϕ dependence, we have just dropped that term. The second derivative term for R_{m4}, for example, is

$$Cm4(\Lambda, k, r, \theta, t) := -\frac{d}{dt}\{[\underline{\Gamma}(\Lambda, k, r, \theta, t)]_{1,1}\}_{m,1} - \frac{d}{dt}\{[\underline{\Gamma}(\Lambda, k, r, \theta, t)]_{2,1}\}_{m,2}$$
$$-\frac{d}{dt}\{[\underline{\Gamma}(\Lambda, k, r, \theta, t)]_{3,1}\}_{m,3} - \frac{d}{dt}\{[\underline{\Gamma}(\Lambda, k, r, \theta, t)]_{4,1}\}_{m,4}$$

The Ricci tensor is then obtained by evaluating the sum of these three contributions symbolically. For example, the component R_{44} is given by[2]

$$A44(\Lambda, k, r, \theta, t) + B44(\Lambda, k, r, \theta, t) + C44(\Lambda, k, r, \theta, t) \text{ simplify} \to \frac{-3}{\Lambda(t)}\frac{d^2\Lambda(t)}{dt^2}$$

The remaining components of the Ricci tensor are obtained through a judicious re-labeling of the indices, and the differentiation variable in $Cmn(\Lambda, k, r, \theta, t)$. This quite marvelously reproduces the results in Eqs. (B.6), eliminating the tedious algebra of the hand calculation.[3]

Problem 13.3 Carry the proof through from the beginning, and verify that the relation in Eq. (13.49) also holds for the Robertson-Walker metric with $k = 0$ as studied in chapter 11.[4]

---

[2]Again, in these calculations $dt = d(ct)$.

[3]And any mistakes in that algebra!

[4]While the result may be obvious, it is a useful exercise to think through each step again from the beginning.

Solution to Problem 13.3

Let us go through the proof of Eq. (13.49) step by step for $k = 0$. In the LF^3, the covariant divergence of the energy-momentum tensor vanishes

$$\bar{T}^{\mu\nu}{}_{;\nu} = 0$$

This relation is preserved under the coordinate transformation to the global inertial laboratory frame, and hence

$$T^{\mu\nu}{}_{;\nu} = \left[P\, g^{\mu\nu} + (\rho c^2 + P) \frac{u^\mu u^\nu}{c^2} \right]_{;\nu} = 0$$

$$\frac{u^\mu}{c} = (0, 0, 0, 1)$$

Here the fluid is at rest in the global inertial laboratory frame, and both $P(t)$ and $\rho(t)$ are functions of time. It follows as in Eq. (10.43) that the covariant divergence takes the form

$$T^{\mu\nu}{}_{;\nu} = [P\, g^{\mu\nu}]_{;\nu} + \left[(\rho c^2 + P) \frac{u^\mu u^\nu}{c^2} \right]_{;\nu}$$

$$= P g^{\mu\nu}{}_{;\nu} + g^{\mu\nu} P_{;\nu} + \left\{ \frac{1}{\sqrt{-g}} \frac{\partial}{\partial q^\nu} \left[\sqrt{-g}\, (\rho c^2 + P) \frac{u^\mu u^\nu}{c^2} \right] \right\} +$$

$$\Gamma^\mu_{\lambda\nu} (\rho c^2 + P) \frac{u^\lambda u^\nu}{c^2}$$

In the second line, the covariant divergence of the metric tensor in the first term vanishes. The second term in the second line only contributes for $\mu = 4$, in which case it gives

$$g^{44} \frac{dP}{d(ct)} = -\frac{dP}{d(ct)}$$

where we now use the metric in Eqs. (11.8). This equation is identical to Eq. (13.45). The last term in the above vanishes since it is evident from Eqs. (11.13) that there is also no Γ^μ_{44} with this metric

$$\Gamma^\mu_{44} (\rho c^2 + P) \frac{u^4 u^4}{c^2} = 0$$

The term in braces in the above also only contributes for $\mu = 4$. Since the square root of the determinant of the metric is now given by $\sqrt{-g} = \Lambda^3(t)$, this term becomes

$$\frac{1}{\sqrt{-g}} \frac{\partial}{\partial q^4} \left[\sqrt{-g}\, (\rho c^2 + P) \frac{u^4 u^4}{c^2} \right] = \frac{d}{d(ct)} (\rho c^2 + P) + (\rho c^2 + P) \frac{1}{\Lambda^3} \frac{d}{d(ct)} \Lambda^3$$

which is identical to Eq. (13.47). Hence the conservation of the energy-momentum tensor is satisfied identically for $\mu = (x, y, z)$, and for $\mu = 4$ it leads to the following relation

$$g^{44} \frac{dP}{d(ct)} + \frac{d}{d(ct)}(\rho c^2 + P) + (\rho c^2 + P)\frac{1}{\Lambda^3}\frac{d}{d(ct)}\Lambda^3 = \dot{0} \qquad ; \mu = 4$$

The term in dP/dt cancels, with the result that

$$\frac{d\rho}{dt} + 3\left(\frac{\dot{\Lambda}}{\Lambda}\right)\left(\rho + \frac{P}{c^2}\right) = 0$$

This equation is indeed identical to that obtained for $k \neq 0$ in Eq. (13.49). A dynamical result arising from the conservation of the energy-momentum tensor, it relates the rate of change of the mass density and the stretching of the space to the current value of the mass density and the pressure, which is obtained from the mass density through the equation of state.

Problem 13.4 (a) Use the Ricci tensor of Eqs. (7.71) and the metric with $A(r) = 1/(1 - kr^2)$ and $B(r) = 1$ to evaluate the *scalar curvature* of the riemannian space described by the Robertson-Walker metric with $k \neq 0$ and $\Lambda = 1$. Show that it is given by

$$R = R^{\mu}{}_{\mu} = 6k \qquad ; \text{ scalar curvature with } \Lambda = 1$$

Hence conclude that it is a space of *constant curvature*.
  (b) What is the answer when $\Lambda(t) \neq 1$?

Solution to Problem 13.4

If $B = 1$ and $B' = 0$, the Ricci tensor $R_{\mu\nu}$ in Eqs. (7.71) reduces to

$$R_{\mu\nu} = 0 \qquad\qquad ; \mu \neq \nu, \quad B = 1$$

$$R_{rr} = \frac{1}{r}\left(\frac{A'}{A}\right)$$

$$R_{\theta\theta} = 1 - \frac{1}{A} + \frac{r}{2A}\left(\frac{A'}{A}\right)$$

$$R_{\phi\phi} = R_{\theta\theta}\sin^2\theta$$

$$R_{44} = 0$$

With $A(r) = 1/(1 - kr^2)$, the last four equations become

$$R_{rr} = \frac{2k}{1 - kr^2} \qquad\qquad ; A(r) = \frac{1}{1 - kr^2}$$

$$R_{\theta\theta} = 2kr^2$$

$$R_{\phi\phi} = 2kr^2 \sin^2\theta$$

$$R_{44} = 0$$

The metric $g^{\mu\nu}$ in this case is, from Eqs. (7.49),

$$g^{\mu\nu} = \begin{bmatrix} (1 - kr^2) & & & \\ & 1/r^2 & & \\ & & 1/r^2 \sin^2\theta & \\ & & & -1 \end{bmatrix}$$

The scalar curvature is then

$$R = R^{\mu}{}_{\mu} = g^{\mu\nu} R_{\nu\mu} = 6k$$

which is the stated answer. We conclude that this metric represents a space of *constant curvature*.[5]

(b) When $\Lambda(t) \neq 1$, the scalar curvature must be evaluated from the metric in Eqs. (B.2) and the Ricci tensor in Eqs. (B.6). All the elements are there, and we leave their combination as an exercise for the reader.

Problem 13.5 Consider the solution to the Einstein Eqs. (13.50) in the case where $k = 0$ and $\rho c^2 \gg P$.[6] Take the positive root, and show that the previous results for a flat, cold, matter-dominated cosmology are reproduced:

$$\frac{\dot{\Lambda}}{\Lambda} = \frac{2}{3(t - t_0)}$$

$$\rho(t) = \frac{1}{6\pi G} \frac{1}{(t - t_0)^2}$$

$$\rho(t_p) = \frac{3}{8\pi G} H_0^2 \qquad\qquad ; H_0 = \text{Hubble's constant}$$

---

[5] Note that the constant k has dimensions $[l^{-2}]$.

[6] We here correct an obvious misprint in the text; this is the proper inequality.

Solution to Problem 13.5

The Einstein field equations for the Robertson-Walker metric with $k \neq 0$ are given in Eqs. (13.50)

$$\dot{\Lambda}^2 - \frac{8\pi G \rho}{3} \Lambda^2 = -kc^2 \qquad \text{; Einstein equations for } k \neq 0$$

$$\dot{\rho} + 3 \left(\frac{\dot{\Lambda}}{\Lambda} \right) \left(\rho + \frac{P}{c^2} \right) = 0$$

Specialize to $k = 0$ in the first equation, and take the positive square root. Assume a matter-dominated universe, with $\rho c^2 \gg P$, in the second. These equations then become

$$\frac{1}{c^2} \left(\frac{\dot{\Lambda}}{\Lambda} \right) = \left(\frac{\kappa}{3} \right)^{1/2} \rho^{1/2}$$

$$\dot{\rho} + 3 \left(\frac{\dot{\Lambda}}{\Lambda} \right) \rho = 0$$

where we have introduced, from Eq. (10.55),

$$\kappa \equiv \frac{8\pi G}{c^4}$$

Now take the time derivative of the first equation

$$\frac{1}{c^2} \left(\frac{\ddot{\Lambda}}{\Lambda} \right) = \left(\frac{\kappa}{3} \right)^{1/2} \rho^{1/2} \left(\frac{\dot{\Lambda}}{\Lambda} \right) + \left(\frac{\kappa}{3} \right)^{1/2} \frac{\rho^{1/2}}{2} \left(\frac{\dot{\rho}}{\rho} \right)$$

Re-substitute the above relations

$$\frac{1}{c^2} \left(\frac{\ddot{\Lambda}}{\Lambda} \right) = \frac{\kappa}{3} \rho c^2 - \frac{\kappa}{2} \rho c^2 = -\frac{\kappa}{6} \rho c^2$$

It follows that

$$\frac{1}{c^2} \left[\left(\frac{\ddot{\Lambda}}{\Lambda} \right) + 2 \left(\frac{\dot{\Lambda}}{\Lambda} \right)^2 \right] = -\frac{\kappa}{6} \rho c^2 + \frac{2\kappa}{3} \rho c^2 = \frac{\kappa}{2} \rho c^2$$

These are the Einstein field Eqs. (11.27)

$$-\frac{3}{c^2} \frac{\ddot{\Lambda}}{\Lambda} = \frac{\kappa}{2} \rho c^2$$

$$\frac{1}{c^2} \left[\frac{\ddot{\Lambda}}{\Lambda} + 2 \left(\frac{\dot{\Lambda}}{\Lambda} \right)^2 \right] = \frac{\kappa}{2} \rho c^2 \qquad \text{; Einstein field equations}$$

which form the basis for our discussion of cosmology in chapter 11, and lead to the stated results for a flat, cold, matter-dominated cosmology.[7]

Problem 13.6 Consider the solution to the Einstein Eqs. (13.50) in the case where the source is simply that of the cosmological constant, with the effective equation of state in Eq. (13.19).

(a) Show that $\rho = \rho_0$ where ρ_0 is a constant. Assume for the purposes of this problem that ρ_0 is positive.

(b) Assume that k/Λ^2 is negligible. Take the positive root, and show the solution is

$$\Lambda \propto e^{\mathcal{H}_0 t} \qquad ; \mathcal{H}_0 \equiv \left(\frac{8\pi G \rho_0}{3} \right)^{1/2}$$

where $\mathcal{H}_0$ is a constant. Hence conclude that in this case Λ exhibits *exponential growth*.

(c) Show that now

$$\frac{k}{(\dot{\Lambda})^2} \propto \frac{k}{\mathcal{H}_0^2} e^{-2\mathcal{H}_0 t}$$

Thus verify that the assumption in (b), which is equivalent to setting $k = 0$ at the outset, is consistent. In this case one *converges with time to the flat solution with* $\Omega = 1$.

Solution to Problem 13.6

The Einstein field Eqs. (13.50) for the Robertson-Walker metric with $k \neq 0$ are

$$\dot{\Lambda}^2 - \frac{8\pi G \rho}{3} \Lambda^2 = -kc^2 \qquad ; \text{Einstein equations for } k \neq 0$$

$$\dot{\rho} + 3 \left(\frac{\dot{\Lambda}}{\Lambda} \right) \left(\rho + \frac{P}{c^2} \right) = 0$$

The effective equation of state (EOS) with a source which is simply that of the cosmological constant is given in Eqs. (13.19)

$$\left(P + \rho c^2 \right)^{\text{vac}} = 0 \qquad ; \text{effective EOS}$$

$$\frac{\bar{\Lambda} c^4}{8\pi G} = P^{\text{vac}}$$

---

[7]Readers can convince themselves that the time derivative of the first relation, and substitution of the second, lead directly to Eqs. (11.27) with no square-root involved.

(a) It follows from the EOS, and the second field equation, that $\rho^{\text{vac}} = \rho_0$ is a constant

$$\rho_0 c^2 = -P^{\text{vac}} = -\frac{\bar{\Lambda} c^4}{8\pi G}$$

As stated, we assume for the purposes of this problem that ρ_0 is positive ($\bar{\Lambda} < 0$).

(b) Assume to start with that $k/\dot{\Lambda}^2$ is negligible. As in Prob. 13.5, take the positive root in the first field equation to obtain[8]

$$\frac{1}{\Lambda(t)} \frac{d\Lambda(t)}{dt} = \left(\frac{8\pi G \rho_0}{3}\right)^{1/2}$$

The solution to this differential equation for the spatial scale factor in the metric is

$$\Lambda(t) \propto e^{\mathcal{H}_0 t} \qquad ; \ \mathcal{H}_0 \equiv \left(\frac{8\pi G \rho_0}{3}\right)^{1/2}$$

where $\mathcal{H}_0$ is a constant.[9] Hence we conclude that in this case $\Lambda(t)$ exhibits *exponential growth*.

(c) One is now in a position to examine the neglected term

$$\frac{k}{(\dot{\Lambda})^2} \propto \frac{k}{\mathcal{H}_0^2} e^{-2\mathcal{H}_0 t}$$

This is exponentially small, verifying the consistency of the starting assumption in (b), which is equivalent to setting $k = 0$ at the outset. In this case one converges with time to the flat solution with $\Omega = 1$ in the Friedmann Eq. (13.41).[10]

Problem 13.7 Go to the LF^3 and introduce cartesian coordinates $x^\mu = (\vec{x}, ct)$. The metric $\bar{g}^{\mu\nu}$ is just that of Minkowski space in Eq. (13.27), so the volume element is $d\tau = d^3x\, d(ct)$. Assume the usual form of the potential in Eq. (13.54), and write the lagrangian density for a scalar field ϕ with inverse Compton wavelength $m = m_0 c/\hbar$ as

$$\mathcal{L}_\phi\left(\phi, \frac{\partial \phi}{\partial x^\mu}\right) = -\frac{c^2}{2}\left[\bar{g}^{\mu\nu}\frac{\partial \phi}{\partial x^\mu}\frac{\partial \phi}{\partial x^\nu} + m^2\phi^2\right]$$

---

[8]We effectively impose the initial condition that $\dot{\Lambda}(0)/\Lambda(0) > 0$.

[9]From Eq. (11.41), $\mathcal{H}_0$ is Hubble's constant in this cosmology.

[10]Note that here $\ddot{\Lambda}/\Lambda = \mathcal{H}_0^2$, which is the basic idea behind "dark energy" leading to an *accelerating* expansion of the universe.

(a) Start from Hamilton's principle in Eq. (13.6) and derive the continuum form of Lagrange's equation[11]

$$\frac{\partial}{\partial x^\mu} \frac{\partial \mathcal{L}_\phi}{\partial(\partial \phi/\partial x^\mu)} - \frac{\partial \mathcal{L}_\phi}{\partial \phi} = 0$$

Here $\partial/\partial x^\mu$ implies that the other coordinates are kept fixed in the differentiation.

(b) Show that the equation of motion of the scalar field is then

$$\left(\Box - m^2 \right) \phi = 0$$

which is the Klein-Gordon equation [compare the second of Eqs. (10.91)].

(c) Now make a coordinate transformation to coordinates q^μ in the global, inertial laboratory frame, assume a general potential $\mathcal{V}(\phi)$, and remember that the scalar field ϕ is invariant. Show the lagrangian density takes the form in Eq. (13.20).

Solution to Problem 13.7

This problem concerns elementary scalar field theory. As stated, we start in the LF^3 where one has just special relativity, and introduce cartesian coordinates $x^\mu = (\vec{x}, ct)$. The metric $\bar{g}^{\mu\nu}$ is just that of Minkowski space in Eq. (13.27), so the volume element is $d\tau = d^3x\, d(ct)$. We first assume the usual quadratic form of the potential in Eq. (13.54), and write the lagrangian density for a scalar field ϕ with inverse Compton wavelength $m = m_0 c/\hbar$ as

$$\mathcal{L}_\phi\left(\phi, \frac{\partial \phi}{\partial x^\mu}\right) = -\frac{c^2}{2}\left[\bar{g}^{\mu\nu}\frac{\partial \phi}{\partial x^\mu}\frac{\partial \phi}{\partial x^\nu} + m^2\phi^2\right]$$

(a) Hamilton's principle in Eq. (13.6) now states that

$$\delta \int \mathcal{L}_\phi\left(\phi, \frac{\partial \phi}{\partial x^\mu}\right) d^4x = 0 \qquad ;\text{ Hamilton's principle}$$

$$\text{fixed endpoints in time}$$

The variation is carried out by letting

$$\phi(x) \rightarrow \phi(x) + \eta(x)$$
$$\frac{\partial \phi(x)}{\partial x^\mu} \rightarrow \frac{\partial \phi(x)}{\partial x^\mu} + \frac{\partial \eta(x)}{\partial x^\mu}$$

---

[11]See [Fetter and Walecka (2003), p. 129].

where $\eta(x)$ is an arbitrary, infinitesimal, function of space-time. A Taylor series expansion in η then gives, for $\eta \to 0$,

$$\delta \int \mathcal{L}_\phi \left(\phi, \frac{\partial \phi}{\partial x^\mu} \right) d^4 x = \int \left[\eta(x) \frac{\partial \mathcal{L}_\phi}{\partial \phi} + \frac{\partial \eta(x)}{\partial x^\mu} \frac{\partial \mathcal{L}_\phi}{\partial (\partial \phi / \partial x^\mu)} \right] d^4 x$$

Here the partial derivatives of $\mathcal{L}_\phi$ on the r.h.s. imply that the other variables in $\mathcal{L}_\phi$ on the l.h.s. are to be held fixed. Now carry out a partial integration of the term in $\partial \eta(x)/\partial x^\mu$. The boundary contributions disappear due to

- Fixed endpoints in the time;
- The spatial boundary conditions:
  - Periodic boundary conditions, *or*
  - A localized disturbance.

The variation of the action then takes the form

$$\delta \int \mathcal{L}_\phi \left(\phi, \frac{\partial \phi}{\partial x^\mu} \right) d^4 x = \int \eta(x) \left[\frac{\partial \mathcal{L}_\phi}{\partial \phi} - \frac{\partial}{\partial x^\mu} \frac{\partial \mathcal{L}_\phi}{\partial (\partial \phi / \partial x^\mu)} \right] d^4 x$$

Since $\eta(x)$ is arbitrary, this implies the continuum form of Lagrange's equation for a scalar field

$$\frac{\partial}{\partial x^\mu} \frac{\partial \mathcal{L}_\phi}{\partial (\partial \phi / \partial x^\mu)} - \frac{\partial \mathcal{L}_\phi}{\partial \phi} = 0 \qquad ; \text{ Lagrange's eqn}$$

(b) If this is applied to the given lagrangian, one finds

$$-\bar{g}^{\mu\nu} \frac{\partial}{\partial x^\mu} \frac{\partial \phi}{\partial x^\nu} + m^2 \phi = 0$$

Since

$$\bar{g}^{\mu\nu} \frac{\partial}{\partial x^\mu} \frac{\partial}{\partial x^\nu} = \nabla^2 - \frac{1}{c^2} \frac{\partial^2}{\partial t^2} = \square$$

is just the wave operator, we obtain the Klein-Gordon equation for the scalar field

$$(\square - m^2)\phi = 0$$

(c) One can now make a coordinate transformation to coordinates q^μ in the global, inertial laboratory frame, assuming a more general potential $\mathcal{V}(\phi)$, where the scalar field ϕ is invariant. The lagrangian density then takes the form in Eq. (13.20)

$$\mathcal{L}_\phi = \left[-\frac{c^2}{2} g^{\mu\nu} \frac{\partial \phi}{\partial q^\mu} \frac{\partial \phi}{\partial q^\nu} - \mathcal{V}(\phi) \right]$$

Problem 13.8 Start from Hamilton's principle in Eq. (13.6) and the scalar field lagrangian of Eq. (13.20).

(a) Show Lagrange's equation for the scalar field takes the form

$$\frac{c^2}{\sqrt{-g}} \frac{\partial}{\partial q^\mu} \left(\sqrt{-g} \, g^{\mu\nu} \frac{\partial \phi}{\partial q^\nu} \right) = \frac{\partial V(\phi)}{\partial \phi}$$

or;
$$c^2 (\nabla^\mu \phi)_{;\mu} = \frac{\partial V(\phi)}{\partial \phi}$$

(b) Assume a spatially constant scalar field which is a function of time $\phi(t)$. Show the equation of motion of the scalar field in the LF^3 is

$$\ddot{\phi} = -\frac{\partial V(\phi)}{\partial \phi}$$

Here $\ddot{\phi} = d^2\phi/dt^2$.

(c) Assume a spatially constant scalar field which is a function of time $\phi(t)$ and the Robertson-Walker metric with $k \neq 0$ of Eq. (13.31). Show the equation of motion of the scalar field in the global inertial laboratory frame takes the form

$$\ddot{\phi} + 3 \left(\frac{\dot{\Lambda}}{\Lambda} \right) \dot{\phi} = -\frac{\partial V(\phi)}{\partial \phi}$$

Here $\dot{\phi} = d\phi/dt$ and $\ddot{\phi} = d^2\phi/dt^2$.

(d) Use the equation of state for the scalar field in Eqs. (13.29) to show that under the stated conditions, the result in part (c) is identical to the second of Eqs. (13.50).

Solution to Problem 13.8

(a) The scalar lagrangian density in Eq. (13.20) is

$$\mathcal{L}_\phi = \left[-\frac{c^2}{2} g^{\mu\nu} \frac{\partial \phi}{\partial q^\mu} \frac{\partial \phi}{\partial q^\nu} - V(\phi) \right]$$

Apply Hamilton's principle in Eq. (13.6) to this lagrangian

$$\delta \int \mathcal{L}_\phi \sqrt{-g} \, dq^1 dq^2 dq^3 dq^4 = 0 \qquad \text{; Hamilton's principle}$$

$$\text{fixed endpoints in time}$$

A repetition of the analysis in Prob. 13.7 leads to Lagrange's equation

$$\frac{\partial}{\partial q^\mu} \frac{\partial(\mathcal{L}_\phi \sqrt{-g})}{\partial(\partial\phi/\partial q^\mu)} - \frac{\partial(\mathcal{L}_\phi \sqrt{-g})}{\partial\phi} = 0 \quad ; \text{ Lagrange's eqn}$$

The Robertson-Walker metric with $k \neq 0$ in Eq. (13.31) gives

$$\sqrt{-g} = \Lambda^3(t) \frac{r^2 \sin\theta}{(1 - kr^2)^{1/2}}$$

Lagrange's equation for the scalar field then reads

$$\frac{\partial}{\partial q^\mu} \left[-c^2 g^{\mu\nu} \frac{\partial\phi}{\partial q^\nu} \sqrt{-g} \right] + \frac{\partial\mathcal{V}(\phi)}{\partial\phi} \sqrt{-g} = 0$$

This is re-written as

$$\frac{c^2}{\sqrt{-g}} \frac{\partial}{\partial q^\mu} \left(\sqrt{-g}\, g^{\mu\nu} \frac{\partial\phi}{\partial q^\nu} \right) = \frac{\partial\mathcal{V}(\phi)}{\partial\phi}$$

$$\text{or;} \qquad c^2 (\nabla^\mu\phi)_{;\,\mu} = \frac{\partial\mathcal{V}(\phi)}{\partial\phi}$$

where the last line follows from the covariant divergence in Eq. (10.31).

(b) Assume a spatially constant scalar field which is only a function of time $\phi(t)$. The equation of motion in the LF^3, where one has just flat Minkowski space with $q^\mu = (x, y, z, ct)$ and the Lorentz metric of Eqs. (6.5)–(6.7), reduces to

$$\ddot{\phi} = -\frac{\partial\mathcal{V}(\phi)}{\partial\phi}$$

Here $\ddot{\phi} = d^2\phi/dt^2$.

(c) The corresponding equation of motion of the scalar field, which is only a function of time $\phi(t)$, in the global inertial laboratory frame follows from part (a) as

$$\ddot{\phi} + 3 \left(\frac{\dot{\Lambda}}{\Lambda} \right) \dot{\phi} = -\frac{\partial\mathcal{V}(\phi)}{\partial\phi}$$

Here $\dot{\phi} = d\phi/dt$ and $\ddot{\phi} = d^2\phi/dt^2$.

(d) The equation of state in Eqs. (13.29) for the scalar field $\phi(t)$ is

$$\rho c^2 = \frac{1}{2}\dot{\phi}^2 + \mathcal{V}(\phi) \qquad ; \text{ EOS for scalar field in } LF^3$$

$$P = \frac{1}{2}\dot{\phi}^2 - \mathcal{V}(\phi)$$

Note that $(\rho, P, \mathcal{V})$ are all scalars under transformation to the global inertial laboratory frame, and hence so is $\dot{\phi}^2$; recall that with the Robertson-Walker metric, the local times intervals dt at a given point are identical.[12]

Now work in the laboratory frame. Differentiate the first relation w.r.t. t, and multiply the result in part (c) by $\dot{\phi}$, to obtain the two equations

$$\dot{\rho}c^2 = \dot{\phi}\ddot{\phi} + \frac{\partial \mathcal{V}(\phi)}{\partial \phi}\dot{\phi}$$

$$\dot{\phi}\ddot{\phi} + 3\left(\frac{\dot{\Lambda}}{\Lambda}\right)\dot{\phi}^2 = -\frac{\partial \mathcal{V}(\phi)}{\partial \phi}\dot{\phi}$$

Add these two relations to obtain

$$\dot{\rho}c^2 + 3\left(\frac{\dot{\Lambda}}{\Lambda}\right)\dot{\phi}^2 = 0$$

Division by c^2, and addition of the two relations in the EOS, gives

$$\dot{\rho} + 3\left(\frac{\dot{\Lambda}}{\Lambda}\right)\left(\rho + \frac{P}{c^2}\right) = 0$$

This is the second of Eqs. (13.50), obtained from the covariant divergence of the energy-momentum tensor.

Problem 13.9 Try to find an approximate solution to Eqs. (13.50) in the case where there is only an additional spatially constant scalar field $\phi(t)$ that satisfies the results in Prob. 13.8. Assume a potential of the form in Fig. 13.2 in the text, and use the equation of state of Eq. (13.29).

(a) Start in the LF^3 where ρ is defined. Differentiate the first of Eqs. (13.29) with respect to time and show that

$$\dot{\rho}c^2 = \left[\ddot{\phi} + \frac{\partial \mathcal{V}(\phi)}{\partial \phi}\right]\dot{\phi} = 0$$

Hence conclude that ρ must be constant everywhere in space as the scalar field develops with time

$$\rho = \frac{1}{2c^2}\dot{\phi}^2 + \frac{1}{c^2}\mathcal{V}(\phi) = \rho_0 \qquad ; \text{ constant}$$

---

[12]Recall Eqs. (13.31). Although the *velocities* $\dot{\phi}$ coincide in the two frames, the *accelerations* $\ddot{\phi}$ do not.

Here $\rho_0 = \mathcal{V}(0)/c^2$, which we define to be $\rho_0 \equiv \mathcal{V}_0/c^2$. Since ϕ is a scalar, and the times in the two frames are identical, observe that this constant mass density also fills all of space in the global inertial laboratory frame.

(b) Substitute the result in (a) into the first of Eqs. (13.50) and show

$$\dot{\Lambda}^2 - \mathcal{H}_0^2 \Lambda^2 = -kc^2 \qquad ; \mathcal{H}_0 \equiv \left(\frac{8\pi G \rho_0}{3}\right)^{1/2}$$

This expression is now identical to that appearing in Prob. 13.6. As done there, assume that $k/\dot{\Lambda}^2$ is negligible. Take the positive root, and show the solution to this differential equation is $\Lambda \propto e^{\mathcal{H}_0 t}$. Thus

$$\frac{k}{(\dot{\Lambda})^2} \propto \frac{k}{\mathcal{H}_0^2} e^{-2\mathcal{H}_0 t} \qquad ; \Lambda \propto e^{\mathcal{H}_0 t}$$

Hence verify the assumption, which is equivalent to setting $k = 0$ at the outset, and conclude that in this case one again *converges with time to the flat solution with* $\Omega = 1$.

(c) It remains to satisfy the second of Eqs. (13.50), which, because of the result in part (a), can only be done approximately. From Eqs. (13.29) one has

$$\left(\rho + \frac{P}{c^2}\right) = \frac{1}{c^2}\dot{\phi}^2$$

If one demands that the last term in the second of Eqs. (13.50) be negligible with respect to, say, the second term in $\dot{\rho}$ in part (a), one should have a consistent approximation. Show this condition becomes (note that $\dot{\phi}^2 = 2[\mathcal{V}_0 - \mathcal{V}(\phi)]$)

$$\left| 6 \left(\frac{\dot{\Lambda}}{\Lambda}\right) \left[\frac{\mathcal{V}_0 - \mathcal{V}(\phi)}{\mathcal{V}'(\phi)\,\dot{\phi}}\right]\right| \ll 1$$

Show that this may be re-written approximately as

$$\left[\frac{48\pi G\,\mathcal{V}_0[\mathcal{V}_0 - \mathcal{V}(\phi)]}{c^2\,[\mathcal{V}'(\phi)]^2}\right]^{1/2} \ll 1$$

Verify that the expression on the l.h.s. is dimensionless.[13] Conclude that this condition is always satisfied in the limit $c^2 \to \infty$. Since the l.h.s. is homogeneous in $\mathcal{V}$, conclude also that this condition is always satisfied in the limit $c^2 \to \infty$ at fixed $\mathcal{V}/c^2$ (that is, at fixed ρ_0).

---

[13] *Hint:* Recall from Prob. 12.7 that the dimensions of G are $[m^{-1}l^3t^{-2}]$, and verify that the dimensions of $\dot{\phi}^2$ are $[ml^{-1}]$. What are the dimensions of $\mathcal{V}$ and $\mathcal{L}$?

Solution to Problem 13.9

(a)The first of Eqs. (13.29) for the equation of state of the spatially uniform scalar field $\phi(t)$ in the LF^3 is

$$\rho c^2 = \frac{1}{2}\dot{\phi}^2 + \mathcal{V}(\phi)$$

Differentiation w.r.t. time then gives

$$\dot{\rho}c^2 = \left[\ddot{\phi} + \frac{\partial \mathcal{V}(\phi)}{\partial \phi}\right]\dot{\phi} = 0$$

Here the last equality follows from the previous result for such a field in Prob. 13.8(b). Hence we conclude that ρ must be a spatially uniform constant as the scalar field $\phi(t)$ develops with time[14]

$$\rho c^2 = \frac{1}{2}\dot{\phi}^2 + \mathcal{V}(\phi) = \rho_0 c^2 \qquad ; \text{ constant}$$

Here $\rho_0 c^2 = \mathcal{V}(0)$, which we define to be $\rho_0 c^2 \equiv \mathcal{V}_0$. Since ϕ is a scalar, and the times in the two frames are identical, we observe that this constant mass density also fills all of space in the global inertial laboratory frame.

(b) Substitute the result in (a) into the first of the Einstein field Eqs. (13.50) for $k \neq 0$. which yields

$$\dot{\Lambda}^2 - \mathcal{H}_0^2 \Lambda^2 = -kc^2 \qquad ; \mathcal{H}_0 \equiv \left(\frac{8\pi G \rho_0}{3}\right)^{1/2}$$

This expression is now identical to that appearing in Prob. 13.6. As done there, we start by assuming that $k/\dot{\Lambda}^2$ is negligible, and again take the positive root. The solution to this differential equation is then $\Lambda \propto e^{\mathcal{H}_0 t}$. Thus

$$\frac{k}{(\dot{\Lambda})^2} \propto \frac{k}{\mathcal{H}_0^2} e^{-2\mathcal{H}_0 t} \qquad ; \Lambda \propto e^{\mathcal{H}_0 t}$$

This ratio is again exponentially small, which justifies setting $k = 0$ at the outset. We conclude that in this case one again *converges with time to the flat solution with* $\Omega = 1$.

(c) The mass density in the global laboratory frame evolves with time due both to the evolution of the scalar field as it slides down the potential hill, and to the stretching of the space as it does so. The first effect, which we here assume to dominate, was dealt with in part (a) and gave $\dot{\rho} = 0$. It then remains to satisfy the second of the field Eqs. (13.50), which, because

---

[14]This is just energy conservation for ρc^2 in the LF^3.

of the result in part (a), can now only be done approximately. From the equation of state (EOS) of the scalar field in Eqs. (13.29) one has

$$\left(\rho + \frac{P}{c^2}\right) = \frac{1}{c^2}\dot{\phi}^2$$

If one demands that the last term in the second of Eqs. (13.50) be negligible with respect to, say, the second term in $\dot{\rho}$ in part (a), one should have a consistent approximation. From part (a)

$$\dot{\phi}^2 = 2[\mathcal{V}_0 - \mathcal{V}(\phi)]$$

Hence the condition that the scalar field effect dominate becomes

$$\left| 6\left(\frac{\dot{\Lambda}}{\Lambda}\right)\left[\frac{\mathcal{V}_0 - \mathcal{V}(\phi)}{\mathcal{V}'(\phi)\dot{\phi}}\right] \right| \ll 1$$

From the first of Eqs. (13.50), the stretching of the space is given by

$$\frac{\dot{\Lambda}}{\Lambda} \approx \left(\frac{8\pi G\rho}{3}\right)^{1/2}$$

With $\rho c^2 \approx \rho_0 c^2 = \mathcal{V}_0$ from (a), the above condition can be rewritten as

$$\left[\frac{48\pi G\,\mathcal{V}_0[\mathcal{V}_0 - \mathcal{V}(\phi)]}{c^2\,[\mathcal{V}'(\phi)]^2}\right]^{1/2} \ll 1$$

As requested, we do some dimensional analysis:[15]

- The dimensions of the lagrangian density and potential energy density are energy/volume

$$[\mathcal{L}] = [\mathcal{V}] = [ml^2t^{-2}l^{-3}] = [ml^{-1}t^{-2}]$$

- The dimensions of the scalar field can be determined from those of $\mathcal{L}$

$$[l^{-2}c^2\phi^2] = [t^{-2}\phi^2] = [\mathcal{L}] = [ml^{-1}t^{-2}]$$
$$\implies \quad [\phi^2] = [ml^{-1}]$$

- The dimensions of G are given in Prob. 12.7

$$[G] = [m^{-1}l^3t^{-2}]$$

---

[15] Here $[m]$ denotes the dimension of mass.

Thus the quantity in the square brackets on the l.h.s. of the above inequality is *dimensionless*

$$[G\phi^2/c^2] = [m^{-1}l^3t^{-2}ml^{-1}t^2l^{-2}] = [1]$$

We conclude that the condition

$$\left[\frac{48\pi G\, V_0[V_0 - V(\phi)]}{c^2\,[V'(\phi)]^2}\right]^{1/2} \ll 1$$

is always satisfied in the limit $c^2 \to \infty$. and since the l.h.s. is homogeneous in V, we also conclude that this condition is satisfied in the limit $c^2 \to \infty$ *at fixed* V/c^2 (that is, at fixed ρ_0).

Problem 13.10 Show through a rescaling of coordinates that to extract the physics of the Robertson-Walker metric it is sufficient to examine the cases with $k = 0$ and $k = \pm 1$.

Solution to Problem 13.10

The Robertson-Walker metric is given in Eqs. (13.31)

$$(ds)^2 = \Lambda(t)^2\left[\frac{(dr)^2}{1 - kr^2} + r^2(d\theta)^2 + r^2\sin^2\theta(d\phi)^2\right] - c^2(dt)^2$$

$$g_{\mu\nu} = \begin{bmatrix} \Lambda^2(t)/(1 - kr^2) & 0 & 0 & 0 \\ 0 & \Lambda^2(t)r^2 & 0 & 0 \\ 0 & 0 & \Lambda^2(t)r^2\sin^2\theta & 0 \\ 0 & 0 & 0 & -1 \end{bmatrix}$$

We are always free to define new generalized coordinates. For $k \neq 0$, go to

$$r' \equiv \sqrt{|k|}\, r$$

At the same time, re-define the scale factor for the spatial coordinates by

$$\Lambda'(t)^2 \equiv \frac{\Lambda(t)^2}{|k|}$$

The new interval, which is physics, is then

$$(ds)^2 = \Lambda'(t)^2\left[\frac{(dr')^2}{1 - r'^2} + r'^2(d\theta)^2 + r'^2\sin^2\theta(d\phi)^2\right] - c^2(dt)^2 \qquad ; k > 0$$

$$= \Lambda'(t)^2\left[\frac{(dr')^2}{1 + r'^2} + r'^2(d\theta)^2 + r'^2\sin^2\theta(d\phi)^2\right] - c^2(dt)^2 \qquad ; k < 0$$

These are just the Robertson-Walker metrics with $k = \pm 1$. Everything follows from this form of the interval. Therefore, in order to extract the physics of the Robertson-Walker metric, it is sufficient to examine the cases with $k = 0$ and $k = \pm 1$.

Problem 13.11 (a) Work in the LF^3. Consider a scalar field $\phi(x)$ with lagrangian density and energy-momentum tensor

$$\mathcal{L}_\phi = -\frac{c^2}{2} \left[\bar{g}^{\mu\nu} \frac{\partial\phi}{\partial x^\mu} \frac{\partial\phi}{\partial x^\nu} + m^2\phi^2 \right]$$

$$\bar{T}_\phi^{\mu\nu} = c^2 \frac{\partial\phi}{\partial x_\mu} \frac{\partial\phi}{\partial x_\nu} + \bar{g}^{\mu\nu} \mathcal{L}_\phi$$

Use the results in Prob. 13.7, and show that the energy-momentum tensor is conserved along the dynamical path

$$\frac{\partial \bar{T}^{\mu\nu}}{\partial x^\nu} = 0 \qquad ; \text{ along dynamical path}$$

(b) Consider a more general form of the lagrangian density $\mathcal{L}(\phi, \partial\phi/\partial x^\mu)$. Show that if the energy-momentum tensor is defined by

$$\bar{T}_\phi^{\mu\nu} = \bar{g}^{\mu\nu} \mathcal{L}_\phi - \frac{\partial\phi}{\partial x_\mu} \frac{\partial\mathcal{L}_\phi}{\partial(\partial\phi/\partial x^\nu)}$$

then it is again conserved along the dynamical path. Discuss the relevance to Eqs. (13.20) and (13.24).

Solution to Problem 13.11

We do part (b) first, as part (a) is then just a special case.

(b) Work in flat Minkowki space.   Given a lagrangian density $\mathcal{L}_\phi (\phi, \partial\phi/\partial x^\mu)$, one constructs the energy-momentum tensor

$$\bar{T}_\phi^{\mu\nu} = \bar{g}^{\mu\nu} \mathcal{L}_\phi - \frac{\partial\phi}{\partial x_\mu} \frac{\partial\mathcal{L}_\phi}{\partial(\partial\phi/\partial x^\nu)}$$

Consider its four-divergence $\partial \bar{T}_\phi^{\mu\nu}/\partial x^\nu$. It is important to remember two things:

- Partial in the four-divergence means keep the other members of the variable set $x^\mu = (x, y, z, ct)$ fixed;
- It is assumed here that $\bar{T}_\phi^{\mu\nu}$ has no *explicit* dependence on x^μ.

Then

$$\frac{\partial \bar{T}^{\mu\nu}}{\partial x^\nu} = \bar{g}^{\mu\nu}\frac{\partial \mathcal{L}_\phi}{\partial x^\nu} - \frac{\partial}{\partial x^\nu}\left[\frac{\partial \phi}{\partial x_\mu}\frac{\partial \mathcal{L}_\phi}{\partial(\partial\phi/\partial x^\nu)}\right]$$

$$= \frac{\partial \mathcal{L}_\phi}{\partial x_\mu} - \frac{\partial^2 \phi}{\partial x_\mu \partial x^\nu}\frac{\partial \mathcal{L}_\phi}{\partial(\partial\phi/\partial x^\nu)} - \frac{\partial \phi}{\partial x_\mu}\frac{\partial}{\partial x^\nu}\frac{\partial \mathcal{L}_\phi}{\partial(\partial\phi/\partial x^\nu)}$$

Now use Lagrange's equation from Prob. 13.7 on the last term

$$\frac{\partial}{\partial x^\nu}\frac{\partial \mathcal{L}_\phi}{\partial(\partial\phi/\partial x^\nu)} = \frac{\partial \mathcal{L}_\phi}{\partial \phi}$$

This results in

$$\frac{\partial \bar{T}^{\mu\nu}}{\partial x^\nu} = \frac{\partial \mathcal{L}_\phi}{\partial x_\mu} - \frac{\partial^2 \phi}{\partial x_\mu \partial x^\nu}\frac{\partial \mathcal{L}_\phi}{\partial(\partial\phi/\partial x^\nu)} - \frac{\partial \phi}{\partial x_\mu}\frac{\partial \mathcal{L}_\phi}{\partial \phi}$$

The first term follows from our previous remarks as the derivative of an implicit function

$$\frac{\partial}{\partial x_\mu}\mathcal{L}_\phi\left(\phi, \frac{\partial \phi}{\partial x^\nu}\right) = \frac{\partial \mathcal{L}_\phi}{\partial \phi}\frac{\partial \phi}{\partial x_\mu} + \frac{\partial \mathcal{L}_\phi}{\partial(\partial\phi/\partial x^\nu)}\frac{\partial^2 \phi}{\partial x_\mu \partial x^\nu}$$

These two terms evidently cancel the last two terms above, and hence this energy-momentum tensor is divergenceless

$$\frac{\partial \bar{T}^{\mu\nu}}{\partial x^\nu} = 0$$

(a) This part follows from the choice of a particular lagrangian density, and resulting energy-momentum tensor

$$\mathcal{L}_\phi = -\frac{c^2}{2}\left[\bar{g}^{\mu\nu}\frac{\partial \phi}{\partial x^\mu}\frac{\partial \phi}{\partial x^\nu} + m^2\phi^2\right]$$

$$\bar{T}_\phi^{\mu\nu} = c^2\frac{\partial \phi}{\partial x_\mu}\frac{\partial \phi}{\partial x_\nu} + \bar{g}^{\mu\nu}\mathcal{L}_\phi$$

Problem 13.12 Suppose one has the Robertson-Walker metric with $k \neq 0$ and all three sources are simultaneously present: (1) uniform matter mass density with its own equation of state; (2) cosmological constant; (3) uniform, time-dependent scalar field $\phi(t)$ with dynamics governed as in Prob. 13.8.

(a) Write the full set of field equations for this system;

(b) Note how the different source terms contribute:

(c) Discuss how one would go about solving these equations.

Solution to Problem 13.12

(a) The Einstein field Eqs. (13.50) in the global, laboratory frame for the Robertson-Walker metric with $k \neq 0$ are

$$\dot{\Lambda}^2 - \frac{8\pi G\rho}{3}\Lambda^2 = -kc^2 \qquad ;\ \text{Einstein equations}$$

$$\dot{\rho} + 3\left(\frac{\dot{\Lambda}}{\Lambda}\right)\left(\rho + \frac{P}{c^2}\right) = 0$$

The scalar field equation in the same frame is given in Prob. 13.8(c)

$$\ddot{\phi} + 3\left(\frac{\dot{\Lambda}}{\Lambda}\right)\dot{\phi} = -\frac{\partial \mathcal{V}(\phi)}{\partial \phi}$$

(b) The energy-momentum tensor source now receives three independent, additive contributions. The equation of state (EOS) correspondingly also receives three contributions

$$\frac{P}{c^2} = \frac{1}{c^2}\left(P_{\text{mat}} + P_{\text{vac}} + P_\phi\right)$$

$$\rho = \rho_{\text{mat}} + \rho_{\text{vac}} + \rho_\phi$$

The matter EOS is $P_{\text{mat}}(\rho_{\text{mat}})$, the effective EOS for the cosmological constant is given in Eqs. (13.19), and the EOS for the scalar field is displayed in Eqs. (13.29)

$$P_{\text{mat}} = P_{\text{mat}}(\rho_{\text{mat}}) \qquad ;\ \text{matter EOS}$$

$$P_{\text{vac}} = -\rho_{\text{vac}}c^2 = \frac{\bar{\Lambda}c^4}{8\pi G} \qquad ;\ \text{cosmological constant}$$

$$P_\phi = \frac{1}{2}\dot{\phi}^2 - \mathcal{V}(\phi) \qquad ;\ \text{scalar field}$$

$$\rho_\phi c^2 = \frac{1}{2}\dot{\phi}^2 + \mathcal{V}(\phi)$$

The source term now contains contributions from ordinary matter, dark energy, and dark matter. One inputs the cosmological constant $\bar{\Lambda}$, the scalar potential $\mathcal{V}(\phi)$, the gravitational constant G, and the metric parameter k.[16] The unknowns are then $[\Lambda(t),\ \rho_{\text{mat}}(t),\ \phi(t)]$.

(c) To solve the above three coupled, non-linear, second-order differential equations, we can write a finite-difference matrix equation for the

---

[16] See Prob. 13.10.

four-component column vector

$$\underline{Z} = \begin{bmatrix} \Lambda \\ \rho_{\text{mat}} \\ \phi \\ d\phi/dt \end{bmatrix}$$

The equations to be solved can be re-written as[17]

$$\frac{\dot{\Lambda}}{\Lambda} = \left(\frac{8\pi G\rho}{3} - \frac{kc^2}{\Lambda^2} \right)^{1/2}$$

$$\dot{\rho}_{\text{mat}} = -3\left(\rho + \frac{P}{c^2} \right)_{\text{mat}} \left(\frac{8\pi G\rho}{3} - \frac{kc^2}{\Lambda^2} \right)^{1/2}$$

$$\ddot{\phi} = -3\,\dot{\phi}\left(\frac{8\pi G\rho}{3} - \frac{kc^2}{\Lambda^2} \right)^{1/2} - \frac{\partial \mathcal{V}(\phi)}{\partial \phi}$$

The techniques developed in Probs. 7.7 and 10.3–10.5 can now be used to write a finite-difference matrix equation in the form

$$\underline{Z}_{n+1} = \underline{Z}_n + \Delta \underline{\mathcal{D}}_n$$

This can be iterated with Mathcad, starting from an initial condition $\underline{Z}_1$.

Readers can now generate and explore their own full cosmologies, and they are urged to do so.

Problem 13.13 (a) Specialize the energy-momentum tensor of a fluid in Eq. (6.131) to describe the motion of a point mass m located at the position $\vec{x}_p(t)$ through the replacements

$$P \to 0$$

$$\rho(x) \to m\,\delta^{(3)}[\vec{x} - \vec{x}_p(t)]$$

---

[17]In obtaining the second line, we have made use of the following relations

$$\dot{\rho}_\phi = \left(\ddot{\phi} + \frac{\partial \mathcal{V}}{\partial \phi} \right)\left(\frac{\dot{\phi}}{c^2} \right) = -3\left(\frac{\dot{\phi}^2}{c^2} \right)\left(\frac{\dot{\Lambda}}{\Lambda} \right)$$

$$= -3\left(\rho + \frac{P}{c^2} \right)_\phi \left(\frac{8\pi G\rho}{3} - \frac{kc^2}{\Lambda^2} \right)^{1/2}$$

Also $\dot{\rho}_{\text{vac}} = (\rho + P/c^2)_{\text{vac}} = 0$. ρ_ϕ can now be calculated from $\underline{Z}$, and $\rho_{\text{vac}} = -\bar{\Lambda}c^2/8\pi G$. It is assumed here that the argument of the square-root is non-negative.

Show the non-relativistic energy-momentum tensor then takes the following form

$$T_p^{ij} = m\,\delta^{(3)}[\vec{x} - \vec{x}_p(t)]\,\frac{dx_p^i}{dt}\frac{dx_p^j}{dt}$$

$$T_p^{i4} = mc\,\delta^{(3)}[\vec{x} - \vec{x}_p(t)]\,\frac{dx_p^i}{dt}$$

$$T_p^{44} = mc^2\,\delta^{(3)}[\vec{x} - \vec{x}_p(t)]$$

(b) Assume that the combination $(8\pi G/c^4)T_p^{\mu\nu}$ is characterized by a small quantity of $O(h)$. Discuss how these relations might be used to model a source in the discussion of gravitational radiation in chapter 12 and Probs. 12.5-12.7.

(c) How would you improve this model?

Solution to Problem 13.13

(a) The energy-momentum tensor of a fluid is given in Eq. (6.131)

$$T^{\mu\nu}(x) = P(x)g^{\mu\nu} + \left[\rho(x) + \frac{P(x)}{c^2}\right]u^\mu(x)u^\nu(x)$$

Here the local four-velocity is

$$\frac{1}{c}u^\mu(x) = \frac{1}{[1 - \beta^2(x)]^{1/2}}[\vec{\beta}(x), 1] \qquad ; \vec{v} = \vec{\beta}c$$

Assume the pressure is negligible, and assume that the mass density comes from a single particle of mass m located at a position $\vec{x}_p(t)$

$$P \to 0$$
$$\rho(x) \to m\,\delta^{(3)}[\vec{x} - \vec{x}_p(t)]$$

It follows that

$$T^{\mu\nu} = m\,\delta^{(3)}[\vec{x} - \vec{x}_p(t)]u^\mu(x)u^\nu(x)$$

The non-relativistic limit of the four-velocity is

$$\frac{1}{c}u^\mu(x) = [\vec{\beta}(x), 1] \qquad ; \vec{v} = \vec{\beta}c \to 0$$

For a single particle, one can then identify the velocity from the above as

$$c\vec{\beta}(x) = \frac{d\vec{x}_p(t)}{dt}$$

Thus, one can just read off the components of the non-relativistic energy-momentum tensor for a point particle

$$T_p^{ij} = m\,\delta^{(3)}[\vec{x} - \vec{x}_p(t)]\,\frac{dx_p^i}{dt}\frac{dx_p^j}{dt} \qquad ; \text{ point particle (NRL)}$$

$$T_p^{i4} = mc\,\delta^{(3)}[\vec{x} - \vec{x}_p(t)]\,\frac{dx_p^i}{dt}$$

$$T_p^{44} = mc^2\,\delta^{(3)}[\vec{x} - \vec{x}_p(t)]$$

(b) When used as a source for gravitational radiation in Prob. 12.5, one has

$$\Box\,\gamma_{\mu\nu}(\vec{x},t) = -\frac{16\pi G}{c^4}(S_p)_{\mu\nu}(\vec{x},t)$$

$$(S_p)_{\mu\nu} = (T_p)_{\mu\nu} - \frac{1}{2}T_p\,g_{\mu\nu}^0$$

where $T_p = (T_p)_\mu{}^\mu$. As discussed in Prob. 12.5, the source must be sufficiently weak that we can use the linearized expressions everywhere; in particular, we employ $g_{\mu\nu}^0$ in the above and use it to raise and lower indices. The analysis of the gravitational radiation emitted from an accelerated particle (or particles) then proceeds as in Probs. 12.5–12.6.

(c) One obvious improvement is to retain the relativistic four-velocity, but the factor of $(1 - \beta^2)^{-1/2}$ greatly complicates matters. One could also, in principal, retain the pressure P created by these moving particles.

Problem 13.14 Show that one can construct a quantity with the dimension of mass from the fundamental constants $(\hbar, c, G)$, and a corresponding length, in the following manner

$$M_{\text{Planck}} = \left(\frac{\hbar c}{G}\right)^{1/2} \qquad ; \text{ Planck mass}$$

$$\frac{\hbar}{M_{\text{Planck}}\,c} = \left(\frac{\hbar G}{c^3}\right)^{1/2} \qquad \text{Planck length}$$

It is widely believed that the effects of quantum gravity must become important at this distance and corresponding energy scale. Find the numerical values of these quantities.[18]

----

[18] Current experiments probe our understanding of the strong, electromagnetic, and weak interactions to a distance scale of the order of 10^{-16} cm. One should take note of the magnitude of the extrapolation to the Planck length.

Solution to Problem 13.14

In m.k.s. units, the relevant fundamental constants are

$$G = 6.673 \times 10^{-20} \text{ km}^3/\text{kg-sec}^2$$
$$c = 2.998 \times 10^5 \text{ km/sec}$$
$$\hbar = 1.055 \times 10^{-34} \text{ J-sec} = 1.055 \times 10^{-40} \text{ kg-km}^2/\text{sec}$$

It is then evident that $(\hbar c/G)^{1/2}$ has dimensions of $[M]$, and $(\hbar G/c^3)^{1/2}$ has dimensions of $[L]$.

The numerical value of these quantities follows as

$$M_{\text{Planck}} = \left(\frac{\hbar c}{G}\right)^{1/2} = 2.177 \times 10^{-8} \text{ kg} \qquad ; \text{ Planck mass}$$

$$\frac{\hbar}{M_{\text{Planck}}\, c} = \left(\frac{\hbar G}{c^3}\right)^{1/2} = 1.616 \times 10^{-38} \text{ km} \qquad ; \text{ Planck length}$$

In the more familiar units of proton mass and proton Compton wavelength

$$m_p = 1.673 \times 10^{-24} \text{ gm} \qquad ; \quad \frac{\hbar}{m_p c} = 2.103 \times 10^{-16} \text{ m}$$

one has

$$M_{\text{Planck}} = 1.301 \times 10^{19}\, m_p$$
$$\frac{\hbar}{M_{\text{Planck}}\, c} = 7.684 \times 10^{-20}\, \frac{\hbar}{m_p c}$$

Appendix A

Reduction of $g^{\mu\nu} \delta R_{\mu\nu}$ to Covariant Divergences

There are no problems included with this appendix.

Appendix B

Robertson-Walker Metric with $k \neq 0$

There are no problems included with this appendix, but see Prob. 13.1–13.2.

Bibliography

Abraham, R., Marsden, J. E., and Ratiu, J. (1993). *Manifolds, Tensor Analysis, and Applications, 2nd ed.*, Springer, New York

Adler, R., Bazin, M., and Schiffer, M. (1965). *Introduction to General Relativity*, McGraw-Hill, New York

Alford, M., Bowers, J. A., and Rajagopal, K. (2001). *Lecture Notes in Physics* **578**, Springer, Berlin, p. 235

Amore, P. (2000). "Introduction to Inflation," term paper in Physics 786, College of William and Mary, Williamsburg, VA (unpublished)

Amore, P. and Arceo, S. (2006). *Phys. Rev.* **D73**, 083004

Amore, P., and Walecka, J. D., (2013). *Introduction to Modern Physics: Solutions to Problems*, World Scientific Publishing Company, Singapore

Amore, P., and Walecka, J. D., (2014). *Topics in Modern Physics: Solutions to Problems*, World Scientific Publishing Company, Singapore

Amore, P., and Walecka, J. D., (2015). *Advanced Modern Physics: Solutions to Problems*, World Scientific Publishing Company, Singapore

Batiz, Z. (2000). "Black Hole Thermodynamics," term paper in Physics 786, College of William and Mary, Williamsburg, VA (unpublished)

Blau, M. (1994). "Lecture Notes on General Relativity," available at http:// www.unine.ch/phys/ string/lecturesGR.pdf

Brown, G. E. (1994). *Nucl. Phys.* **A574**, 217c

Chandra (2006). *The Chandra X-ray Observatory*, http://chandra.nasa.gov/

Chen, P., and Tajima, T. (1999). "Testing Unruh Radiation with Intense Lasers," *Phys. Rev. Lett.* **83**, 256

COBE (2006). *The Cosmic Background Explorer*, http://lambda.gsfc.nasa.gov/ product/cobe/

Dimopoulos, S., and Susskind, L. (1978). *Phys. Rev.* **D18**, 4500

Einstein, A. (1905). "Zur Electrodynamik Bewegter Körper," *Annalen der Physik* **17**, 891

Einstein, A. (1916). "Die Grundlage der Allgemeinen Relativitäts Theorie," *An-*

nalen der Physik **49**, 50

Einstein, A. (1936). *Science* **84**, 506

Einstein (2006). *Gravity Probe B*, http://einstein.stanford.edu/

Falco, E. E. (1994). *New Journal of Physics* **7**, 200

Fetter, A. L. and Walecka, J. D. (2003). *Theoretical Mechanics of Particles and Continua*, McGraw-Hill, New York (1980); reissued by Dover Publications, Mineola, New York

Fetter, A. L. and Walecka, J. D. (2003a). *Quantum Theory of Many-Particle Systems*, McGraw-Hill, New York (1971); reissued by Dover Publications, Mineola, New York

Foster, J., and Nightengale, J. D. (2006). *A Short Course on General Relativity, 3rd ed.*, Springer, New York

Freedman, W. L., *et al.* (2001). *The Astrophysical Journal* **553**, 47 (2001)

Friedmann, A. (1922). *Z. Phys.* **10**, 377; also **21**, 326 (1924)

Glendenning, N. K. (2000). *Compact Stars: Nuclear Physics, Particle Physics and General Relativity, 2nd ed.*, Springer, New York

Guth, A. H. (2000). "Inflation and Eternal Inflation," *Physics Reports* **333**, 555

Hartle, J. (2002). *Gravity: An Introduction to Einstein's General Relativity*, Addison Wesley, Reading, MA

Hubble (2006). *The Hubble Space Telescope*, http://hubble.nasa.gov/ ; for images, see http://hubblesite.org/gallery/

Hughston, L. P., and Tod, K. P. (1990). *An Introduction to General Relativity*, Cambridge U. Press, New York

Kerr, R. (1963). *Phys. Rev. Lett.* **11**, 237

Kolb, E. W., and Turner, M. S. (1990). *The Early Universe*, Addison-Wesley, Reading, MA

Landau, L. D., and Lifshitz, E. M. (1975). *Classical Theory of Fields*, Pergamon Press, New York

LIGO (2006). *The Laser Interferometer Gravitational-Wave Observatory (LIGO)*, www.ligo.caltech.edu/

Linde, A. (1990). *Particle Physics and Inflationary Cosmology*, Harwood, New York

Misner, C. W., Thorne, K. S., and Wheeler, J. A. (1973). *Gravitation*, W. H. Freeman, San Francisco

Müller, H., and B. D. Serot, B.D. (1996). *Nucl. Phys.* **A606**, 508

Ohanian, H. C., and Ruffini, R. (1994). *Gravitation and Spacetime, 2nd ed.*, W. W. Norton and Company, New York

Ohanian, H. C. (1995). *Modern Physics, 2nd ed.*, Prentice-Hall, Upper Saddle River, NJ

Oppenheimer, J. R., and Volkoff, G. M. (1939). *Phys. Rev.* **55**, 374

Peacock, J. A. (1999). *Cosmological Physics*, Cambridge U. Press, Cambridge, UK

Peebles, P. J. E. (1993). *Principles of Physical Cosmology*, Princeton U. Press, Princeton, NJ

Pound, R. V., and Rebka, G. A. (1960). *Phys. Rev. Lett.* **4**, 337

Robertson, H. P. (1935). *Ap. J.* **82**, 284; also **83**, 187, 257 (1936)

Sandberg, V. (2016). *Colloquium on LIGO*, Thomas Jefferson National Accelerator Facility, October 29, 2016

Schwarzschild, K. (1916). "Über das Gravitationsfeld eines Massenpunktes nach der Einsteinschen Theorie," *Sitzungsberichte der Königlich Preussischen Akademie der Wissenschaften* **1**, 189

Serot, B. D., and Walecka, J. D. (1986). "The Relativistic Nuclear Many-Body Problem," *Advances in Nuclear Physics* **16**, eds. J. W. Negele and E. Vogt, Plenum Press, New York

Shutz, B. F. (1985). *A First Course in General Relativity*, Cambridge U. Press, New York

Sky and Telescope (2000). "A Black Hole for Nearly Every Galaxy," May, 2000, p. 22

Sky and Telescope (2006). "Putting Einstein to the Test," July, 2006, p. 32

Sky and Telescope (2006a). "Einsteinian Energy Bolstered," April, 2006, p. 23

Sky and Telescope (2006b). "Case Strengthened for Inflation," June, 2006, p. 22

Sky and Telescope (2016). "Detection of Gravitational Waves," May, p. 10; October, p. 10

Taylor, E. F., and Wheeler, J. A. (2000). *Exploring Black Holes: Introduction to General Relativity*, Addison-Wesley, Reading, MA

Tolman, R. C. (1939). *Phys. Rev.* **55**, 364

Unruh, W. (1976). *Phys. Rev.* **D14**, 870

Wald, R. M. (1984). *General Relativity*, U. Chicago Press, Chicago

Walecka, J. D. (2004). *Theoretical Nuclear and Subnuclear Physics, 2nd ed.*, World Scientific, Singapore

Walecka, J. D., (2007). *Introduction to General Relativity*, World Scientific Publishing Company, Singapore

Walecka, J. D., (2008). *Introduction to Modern Physics: Theoretical Foundations*, World Scientific Publishing Company, Singapore

Walecka, J. D., (2010). *Advanced Modern Physics: Theoretical Foundations*, World Scientific Publishing Company, Singapore

Walecka, J. D., (2011). *Introduction to Statistical Mechanics*, World Scientific Publishing Company, Singapore

Walecka, J. D., (2013). *Topics in Modern Physics: Theoretical Foundations*, World Scientific Publishing Company, Singapore

Walker, A. G. (1936). *Proc. Lond. Math. Soc.* **42**, 90

Weinberg, S. (1972). *Gravitation and Cosmology: Principles and Applications of the General Theory of Relativity*, John Wiley and Sons, New York

Wiki (2006). *The Wikipedia*, http://en.wikipedia.org/wiki/(topic)

WMAP (2006). *The Wilkinson Microwave Anisotropy Probe (WMAP)*, http://map.gsfc.nasa.gov/

Index

Printed in the United States
By Bookmasters